Building Resilience for Uncertain Water Futures

Patricia Gober

Building Resilience for Uncertain Water Futures

palgrave
macmillan

Patricia Gober
School of Geographical Sciences and Urban Planning
Arizona State University
Tempe, AZ, USA

ISBN 978-3-030-10032-2 ISBN 978-3-319-71234-5 (eBook)
https://doi.org/10.1007/978-3-319-71234-5

Softcover re-print of the Hardcover 1st edition 2018

Cover designed by Getty/artisteer

Printed on acid-free paper

This Palgrave Macmillan imprint is published by the registered company Springer International Publishing AG part of Springer Nature
The registered company address is: Gewerbestrasse 11, 6330 Cham, Switzerland

To Howard, my partner in water science and in life

Preface

This book follows from a graduate seminar in water policy taught in the Schools of Sustainability at Arizona State University in Tempe, Arizona, and Public Policy at the University of Saskatchewan in Saskatoon, Canada. Over the course of ten years, the course attracted an interdisciplinary audience of students from geography, sustainability, civil engineering, public administration, policy studies, and environmental science. Some had considerable knowledge of a narrow field of water science but wanted a broader synthesis of water problems and their potential solutions. They were attracted to the social science and policy dimensions of the course and to the effort to integrate those perspectives with climatology, hydrology, and ecology.

The course evolved somewhat over the ten years to include the perspectives of the international students and more international issues, but the storyline remained constant. It begins with the woeful state of global water resource management, overuse, and increasing pressure on water systems from climate change. Emphasis is on the lack of human capacity to deal with climate change rather than on a warmer planet per se. The water sector's longstanding practices of optimization and prediction, efforts to control water systems for human use, separation of water decisions from public input, and emphasis on water engineering practice present particular problems in the face of today's rapid and uncertain environmental and societal change.

Students were aware of the many vulnerabilities in existing water systems for people, the economy, and environment. They were eager for a more structured way of thinking about vulnerability from a natural hazards perspective. The idea of "hidden vulnerabilities" that lie in water's complex relationships with food, energy, and land and transcend scale particularly intrigued them. Also significant were competing values about water that are manifest in the way cities price water, difficult choices made during drought events, and efforts by Indigenous peoples to reassert their water rights.

At this point in the class, it was clear that today's water issues are not solely the sort of technical and scientific problems seen in coursework in hydrology, policy, and geography. These are complex system problems with the capacity for surprise and problems of deep uncertainty where there are profound value-based disagreements about how to proceed. Considerable discussion centered on the recent climate attitudinal data because they reveal deep societal divides and competing worldviews about climate adaptation and, by implication, water management. Case studies at local and regional levels reveal the different ways that people see water as having value in their lives and the trade-offs they are willing to make to get what they want from the water sector.

The last part of the course shifts to ideas that improve long-term planning and adaptation. Resilience involves the capacity of a system to absorb recurrent natural and human shocks without slowly degrading or even unexpectedly flipping into a less desirable state. Decision Making Under Uncertainty (DMUU) provides a framework for addressing problems of deep uncertainty where parties do not agree on the problem definition itself, the probability distributions that convey risk, and who will be winners and losers in this scheme. Wicked problems are, by definition, complex, uncertain, and controversial. Managing water as a wicked problem requires greater emphasis on the social aspects that surround water problems. Socio-hydrology is a new trend in hydrology that includes the human dimensions in water science and the social processes that connect the science to decision-making and public values. The idea of social learning, in which people learn best in groups, has potential to unite scientists, decision-makers, and the public in efforts to solve local water problems.

Each year, members of the class collaborated to develop a framework for evaluating urban climate adaptation efforts worldwide. These adaptation efforts include water problems (sea level rise, storm surge, flooding, water quality, sanitation, and resources), but they are more broadly focused on adapting cities to climate change. Cities are logical places for policy action because they are at the level where most people experience water problems and understand the trade-offs in decision-making. In a structured evaluation of adaptation efforts at the city scale, students explored water's complicated relationship with land, food, energy, urban design, urban heat islands, and economy. They saw the ideas of resilience, DMUU, wicked water, socio-hydrology, and social learning put into action on the ground across the world. They came away with an appreciation that today's water problems are not simple technical and science problems but social and policy issues that reflect competing values, hidden vulnerabilities, deep uncertainties, and potential for surprise.

This book evolved from class reading lists, projects, guest speakers, discussion groups, movies, and students from across the world who shared their own very personal experiences with water and water problems. It provided me with a way to see my own research about water, climate, and adaptation issues in the Phoenix area and Western Canada as part of a large global story about water resource management in an era of rapid and uncertain environmental and societal change.

Tempe, AZ, USA Patricia Gober

Acknowledgments

This book stems from work supported by the National Science Foundation (NSF) under Grant No. SES-142086, DMUU: DCDCIII: Transformational Solutions for Urban Water Sustainability Transitions in the Colorado River Basin. Any opinions or recommendations expressed in this material are those of the author and do not necessarily reflect the views of the National Science Foundation through the Decision Center for a Desert City (SES-0345945). Support also came from the Canada Excellence Research Chair in Water Security, with funding from Canada's Federal Government National Science and Engineering Council (NSERC), the Province of Saskatchewan and the University of Saskatchewan.

Special thanks to Rebecca Eden for graphic design, Howard Wheater for advice on the manuscript, and Melody English for help obtaining permissions.

Contents

Acronyms

ADWR	Arizona Department of Water Resources
AMA	Active Management Area
BIA	Bureau of Indian Affairs
C	Centigrade
CAD	Canadian dollar
CAGRD	Central Arizona Groundwater Replenishment District
CAP	Central Arizona Project
CBC	Canadian Broadcasting Corporation
CC&Rs	covenants, codes and restrictions
CO_2	carbon dioxide
CSIRO	Commonwealth Scientific and Industrial Research Organization
DCDC	Decision Center for a Desert City
DMUU	Decision Making Under Uncertainty
EQCC	Quito Strategy for Climate Change
EQCC	Quito Strategy for Climate Change
ET	evapotranspiration
FIFA	Fédération Internationale de Football Association, International Federation of Association Football
FITFR	First-In-Time-First-In-Right water allocation system
GHG	Greenhouse Gas
GMA	Groundwater Management Act of 1980
GNP	Gross National Product

GRACE	Gravity Recovery and Climate Experiment
GRIC	Gila River Indian Community
HOA	homeowner association
IPCC AR5	Intergovernmental Panel on Climate Change Assessment Report 5
IPCC	Intergovernmental Panel on Climate Change
MDB	Murray-Darling Basin
MDBA	Murray-Darling Basin Authority
MEC	Metropolitan East Coast
NAFTA	North American Free Trade Agreement
NASA	National Oceanic and Atmospheric Administration
NVCH	North Village Cumberland House
NWI	National Water Initiative
PDSI	Palmer Drought Severity Index
PUB	Public Utilities Board
RO	Reverse Osmosis reclaimed water
SNWA	Southern Nevada Water Authority
SWTP	South-North Water Transfer Project
TGD	Three Gorges Dam
UHI	urban heat island
UKCIP	United Kingdom Climate Impact Program
UNESCO	United Nations Educational, Scientific, and Cultural Organization
UNESCO-IHE	Delft Institute for Water Education
USCRP	United States Climate Research Program
USDA	United States Department of Agriculture
USGU	United States Geological Survey
WELF nexus	water-energy-land-food nexus
WRM	Water Resources Management

List of Boxes

List of Figures

List of Tables

1

Case for Adaptation

1.1 Introduction

Today's world faces existential threats to water systems in the form of intensifying floods and droughts, increasing risks to global supply chains, chronic shortages, hidden vulnerabilities in the water-energy-land-food (WELF) nexus, increasing pollution, and degrading aquatic ecosystems. These threats occur in the context of unsustainable use, environmental change, fragmented and weak governance, and a global economy that is heavily water dependent (World Water Assessment Program 2016). This book makes the case that business-as-usual water science, management, and decision-making institutions are not up to the task of solving today's global water problems and planning for an uncertain future. A path forward using ideas from resilience theory, Decision Making Under Uncertainty (DMUU), socio-hydrology, adaptive management, and social learning offers a new paradigm for water planning and policy.

P. Gober, *Building Resilience for Uncertain Water Futures*,
https://doi.org/10.1007/978-3-319-71234-5_1

1.2 Book Overview

Chapter 2 begins with a discussion of the human dimensions of water security. Only 2.5% of Earth's water is freshwater. The rest is saline and found in the ocean. Most of the freshwater is in glaciers and ice caps (68.7%) and stored as groundwater (30.1% percent). Only 1.2% is surface water available to meet human needs (USGS 2017). Agriculture, industry, and communities use groundwater to supplement surface supplies, but recent evidence from NASA's twin GRACE (Gravity Recovery and Climate Experiment) satellites shows that human consumption is depleting about one-third of Earth's largest groundwater basins with little understanding of how much is left (Richey et al. 2015) (Fig. 1.1).

Worldwide, 884 million people lack access to a basic drinking water service (an improved drinking water source within a round trip of 30 minutes), and 159 million depend upon surface water such as rivers, lakes, ponds, springs, and canals for their main drinking water source. More than two billion people use drinking water contaminated with fecal material; 2.3 billion do not have access to toilets or latrines; and 890 million of them defecate in the open, behind bushes, in gutters, and in open water bodies (United Nations World Health Organization 2017a, b). Eighty percent of the world's population now lives in areas with high risk of either human water insecurity or serious biodiversity loss (Vörösmarty et al. 2010). Water problems are global in scale, take a variety of forms, affect significant numbers of people, and belie easy solutions.

Chapter 2 also shows how societies react to extreme natural events, such as floods, droughts, and sea level rise. Social scientists in the natural hazards field have long warned that planning for and recovering from extreme natural events requires a refocus from the natural events themselves to the human conditions that make people vulnerable to them. Emphasis is on the root causes and dynamic pressures that explain why so many people live in unsafe places, why seemingly predictable events cause so much damage and loss of life, and how system dynamics can lead to unintended consequences and hidden vulnerabilities. After some 75 years of study, hazards research offers keen insight into planning for climate change, starting with reducing the vulnerabilities that already exist such

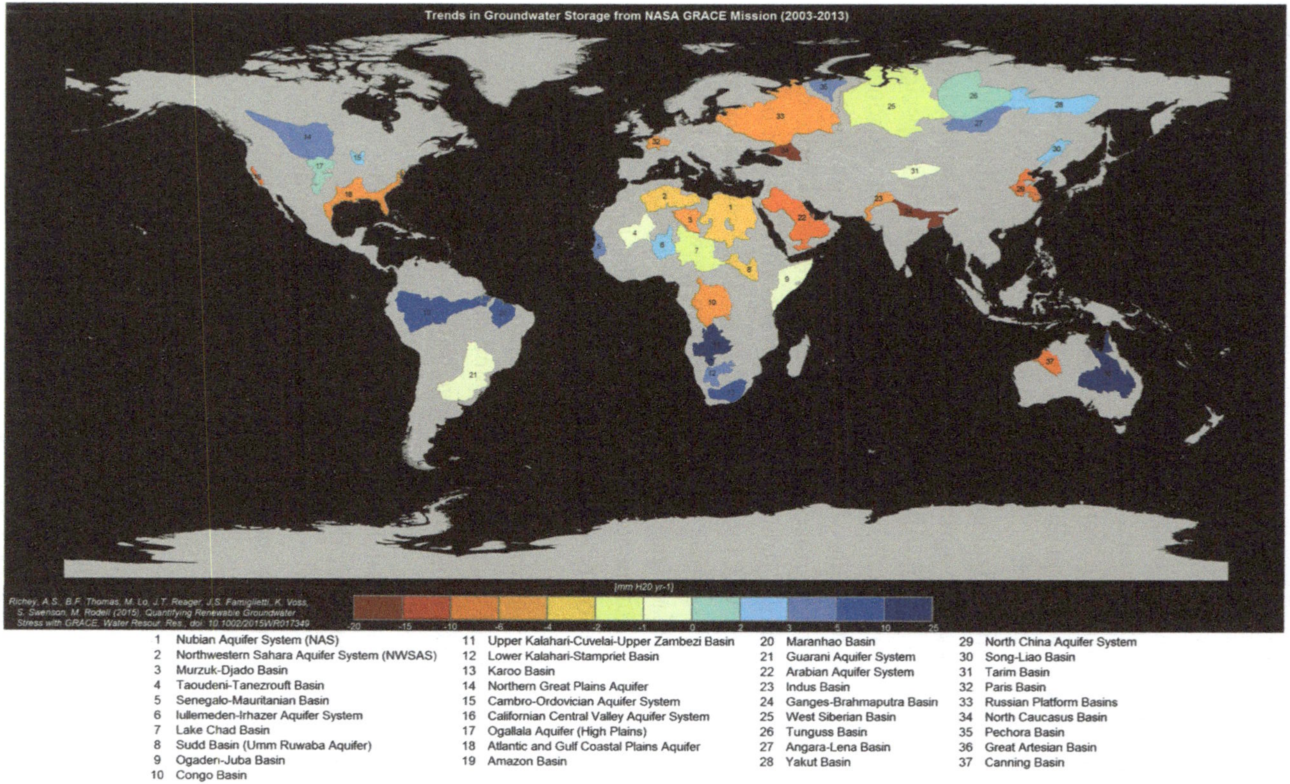

Fig. 1.1 Estimated levels of groundwater stress based on measurements from GRACE satellites. The red aquifers are the most severely stressed by human consumption. UC Irvine/NASA/JPL/Caltech

as moving people out of floodplains, protecting water systems from contamination and salinization, observation and monitoring, building infrastructure to deal with unexpected change, conservation as precautionary action, and improving governance.

Despite large investments in engineering works for water conveyance, storage, and treatment, North America and Europe experience infrastructure failure, water quality problems, loss of ecological diversity, and sea level rise. In 2015, the State of California, one of the world's most productive agricultural regions, lost an estimated $1.84 billion and more than 10,000 jobs in the agricultural sector during recent severe drought conditions (Kerlin 2015). Also significant was the loss of fish species and habitats crucial for sustaining biodiversity as state regulators sought to sustain agricultural production in the face of severe and long-lasting drought. The western Canadian city of Calgary reported more than $6 billion (CAD) in flood damages from an event with a return period of two years per century (City of Calgary 2014). The Aral Sea has all but disappeared due to overuse in a Soviet plan to grow cotton in Central Asia in the 1960s. Case studies of the California drought, Calgary floods, and Aral Sea environmental disaster offer insight into the failure of human agency to prepare for an uncertain future.

Superimposed on this global water landscape punctuated by severe water stress and vulnerability to hazard, future climate change threatens standard methods of water management and planning. Water is the main mechanism for delivering climate change impacts to human populations (Jiménez Cisneros et al. 2014, p. 234). After some 30 years of intensive study, Intergovernmental Panel on Climate Change (IPCC) scientists agree that: "Warming of the climate system is unequivocal, and many observed changes since the 1950s are unprecedented over decades to millennia. The atmosphere and ocean have warmed, the amounts of snow and ice have diminished, sea level has risen, and the concentrations of greenhouse gases have increased" (Stocker et al. 2013). Despite this unwavering, full-throated statement about the onset of global warming, scientists express profound uncertainties about the direction, severity, and geographic distribution of climate change impacts on regional and local water systems. Chapter 3 presents sources of uncertainty regarding climate change impacts and other factors that affect the water sector including demand, technology, regulation, economy, and politics. It argues that these uncertainties present daunting challenges for local and regional water management and planning as practiced today. Traditional methods of decision analysis based on risk assessments, cost-benefit studies, and optimization modeling fail to incorporate the deep uncertainties about climate change and the institutional change needed to cope with them.

There is growing concern that assessments focused on water resources alone ignore vulnerabilities related to water's connections to energy, land, and food, the so-called WELF nexus. Chapter 4 tackles these interconnections. The international trade in food includes embedded water sometimes with unintended consequences. The 2010 Russian drought and heat wave, for example, hurt domestic wheat production, raised global food prices, and reduced access to food by poor people. It was one aspect of the social unrest that led to the Arab Spring (Perez 2013). Future climate conditions will stress energy production infrastructure in all US regions—particularly those with the most water-intensive generation portfolios (US Department of Energy 2013). A major threat to the US energy grid is the localized lack of water for cooling thermoelectric power plants (US Department of Energy 2014). Increased air and water temperatures will increase the likelihood of partial or full shutdown of generation facilities. The so-called governance gap between land use planning and water in the American West allows land use planners to assume that water will be available for projected growth and will not limit development (Bates 2012). In 2016, the Arizona State Legislature passed bills that would have made it possible for developers to construct new homes in places that do not have sufficient groundwater to meet the state's 100-year assured supply rule without informing new homeowners of the water status of their properties. The state's governor vetoed the bills, but they came dangerously close to realizing Bates' gap between land and water management (Howard Fisher Media Services 2016; Glennon and Leshy 2016). Reducing hidden vulnerabilities in water systems increasingly involves the capacity to manage and model WELF relationships (Moss et al. 2016).

Water policies develop in a cultural context where human beliefs, attitudes, and values influence how society processes and responds to climate- and water-related hazards. Chapter 5 lays out the cultural context and values basis for climate and water policy. Water is value laden, locally significant, and non-substitutable. Communities and countries have vastly different views of water as a human right, water as a commodity, and the role of the private sector versus the government in resource allocation and regulation. Recent public opinion surveys provide insight into global attitudes about climate change and the science basis for them. This research shows remarkable international variation in public acceptance of climate science and willingness to adopt policies to mitigate global warming. Despite widespread consensus in the international scientific community as reflected in IPCC reports, sizable segments of the global population do not believe that climate change will affect them personally or, if it does, there are other more serious threats to their personal security such as terrorism or economic insecurity. In-depth analysis of US attitudes shows how polarized public opinion is about this issue. *Global Warming's Six Americas* reflect how different subsets of the population (the six Americas) perceive the issue of climate change—how they vary in level of concern, how much uncertainty there is on the issue, the role of human activity in causing it, potential harm to people and the environment, and trust in scientists (Roser-Renouf et al. 2014). This chapter sets forth the divergent public opinions that have limited global climate policy making and the differing values about water that complicate water management and planning.

Chapter 6 acknowledges that contemporary water problems are uncertain, complex, and riddled with controversy. These types of problems are well suited for Decision Making Under Uncertainty (DMUU) strategies. DMUU moves water planning and management away from the normative, rational approach of optimizing profits and maximizing efficiencies to incorporate differing views about the meaning, purpose, and value of water. Emphasis is on managing trade-offs across sectors and interest groups and planning for an uncertain future. The Australians in particular have embraced this approach to how water problems are framed, the modeling tools required to support decision-making, and how users and scientists come together to solve uncertain water problems (Head 2010; Keim et al. 2016). DMUU redefines the water research question from how to manage risk and maximize returns to how to build adaptive capacity into human engineered water systems and societal institutions.

Emphasis is on integrating multiple sources of knowledge, designing methods to represent alternative possible futures (scenarios), assessing the consequences of particular policy decisions, and including stakeholder viewpoints in model design, implementation, and communication.

Chapter 7 focuses on urban climate adaptation as an exercise in DMUU. Cities around the world are bracing for climate change, many with plans that reflect deep uncertainty, foresight, scenarios, and the need for public buy-in. Adaptation plans do not necessarily focus on water, but because of the interconnectedness of urban resource systems, they engage the water sector in myriad ways.

Chapter 8 tackles the social capacity to deal with water as a wicked problem. Until recently, the mission of water agencies has been relatively clear—to deliver clean water needed to meet a community's existing and growing domestic and economic water needs reliably and at a reasonable cost. Not hearing from the public was tantamount to success in this context, and water managers typically avoided public scrutiny (Lach et al. 2005). The water resource management profession evolved over almost 200 years with great discretion about how to define water problems and design strategies to solve them. Today's water problems reflect increased pressure on a fixed resource, deep uncertainties about environmental and societal change, and a new awareness that decision-making involves making difficult political choices. Social learning supports DMUU in various parts of the world. The idea of knowledge mobilization incorporates problems of and best practices in translating scientific results into tools for decision-making. Citizen science addresses the need for public debate about science-based water issues. It also includes what scientists can learn from citizens about how to frame science questions and how to build public support for science-based decision-making. Stakeholder engagement ensures that management decisions reflect the wide range of human beliefs, attitudes, and values about the problem at hand. The end goal is to marry scientific practice, public engagement, and decision-making institutions into a process of social learning.

Chapter 9 concludes by drawing attention to the need for paradigm change in the water sector. Managing for resilience moves away from controlling water systems and toward making them more flexible for an uncertain future. Water science needs to work in collaboration with managers to monitor ongoing change, assess the consequences of management decisions on system dynamics, and design a future that is robust to a range of future climate conditions. Stakeholder engagement needs to move from one-off input to an ongoing process of building ownership and accountability. Social learning is at the crux of scientific discovery, decision-making, and public participation in managing water problems.

Recognizing that water is local, this book uses case studies from across the world to demonstrate strategies to prepare the water sector for the uncertainties of climate change. Each community grapples with the choices and trade-offs relevant to its unique set of environmental and social circumstances. Local narratives provide insight into place-based water issues in the hope that adaptation strategies can be improved and diffused through networks to like-minded communities elsewhere. They offer a menu of experimental strategies to navigate paradigm change from control to building resilience for an uncertain future.

References

Bates, Sarah. 2012. "Bridging the Governance Gap: Emerging Strategies to Integrate Water and Land Use Planning." *Natural Resources Journal* 53(1): 61–97.

City of Calgary. 2014. "*2013 River Flooding: A City of Calgary Perspective*." Accessed September 12 2017. http://www.cwra-ab-events.org/Presentations/2014/S3-Frigo.pdf.

Glennon, Robert and John Leshy. "Our Turn: 2 Bills Undermine Arizona's Water Supply," Arizona Republic (Phoenix AZ), February 23, 2016. http://www.azcentral.com/story/opinion/op-ed/2016/02/23/glennon-leshy-arizona-legislation-undermines-water-security/80608728/. Accessed August 22 2017.

Head, Brian. 2010. "Wicked Problems in Water Governance: Paradigm Changes to Promote Water Sustainability and Address Planning Uncertainty." Urban Water Security Technical Alliance, Technical Report No. 38. http://www.urbanwateralliance.org.au/publications/UWSRA-tr38.pdf. Accessed September 12 2017.

Howard Fisher Capital Media Services. "Gov. Ducey Vetoes Two Measures Easing Water Rules for Developers." May 9, 2016. http://tucson.com/news/

local/gov-ducey-vetoes-two-measures-easing-water-rules-for-developers/article_fe087d0c-161c-11e6-8789-f328378d6560.html. Accessed September 12 2017.

Jiménez Cisneros, B. E., T. Oki, N. W. Arnell, N. W. G. Benito, J. G. Cogley, P. Döll, T. Jiang, and S. W. Mwakalila. 2014. *Freshwater Resources*. In: *Climate Change 2014: Impacts, Adaptation, and Vulnerability. Part A: Global and Sectoral Aspects*. Contribution of Working Group II to the Fifth Assessment Report of the Intergovernmental Panel on Climate Change, edited by C. B. Field, V. R. Barros, D. J. Dokken, K. J. Mach, M. D. Mastrandrea, T. E. Bilir, M. Chatterjee, K. L. Ebi, Y. O. Estrada, R. C. Genova, B. Girma, E. S. Kissel, A. N. Levy, S. MacCracken, P. R. Mastrandrea, and L. L. White, 229–269. Cambridge, UK and New York: Cambridge University Press.

Keim, Anthony S., Emma K. Austin, and Danielle C. Verdon-Kidd. 2016. "Water Resource Management in a Variable and Changing Climate: Hypothetical Case Study to Explore Decision Making under Uncertainty." *Journal of Water and Climate Change*, 7(2). https://doi.org/10.2166/wcc.2015.040.

Kerlin, Kat. 2015. "Drought Costs California Agriculture $1.48B and 10,100 Jobs in 2015," *Food & Agriculture*. https://www.ucdavis.edu/news/drought-costs-california-agriculture-184b-and-10100-jobs-2015/. Accessed September 12 2017.

Lach, Denise H., Helen Ingram, and Steve Rayner. 2005. "Maintaining the Status Quo: How Institutional Norms and Practices Create Conservative Water Organizations." *Texas Law Review* 83(7): 2027–2053.

Moss, Richard H., KarenFisher-Vanden, Alison Delgado, Scott Backhaus, Christopher L. Barrett, Bhaduri, Budhendra, Ian P. Kracunas, Patrick M. Reed, Jennie S. Rice, Ian Sue Wing, and Claudia Tebaldi. 2016. "Understanding Dynamics and Resilience in Complex Interdependent Systems: Prospects for a Multi-Model Framework and Community of Practice." Report of a workshop held under the auspices of the US Global Change Research Program. Accessed September 12 2017. https://climatemodeling.science.energy.gov/publications/understanding-dynamics-and-resilience-complex-interdependent-systems.

Perez, Ines. "Climate Change and Rising Food Prices Heightened Arab Spring." *Scientific American*, March 4, 2013. https://www.scientificamerican.com/article/climate-change-and-rising-food-prices-heightened-arab-spring/. Accessed on August 11 2017.

Richey, Alexandra S., Brian F. Thomas, Min-Hui Lo, John T. Reager, James S. Famiglietti, Katalyn Voss, Sean Swenson, and Matthew Rodell. 2015.

"Quantifying Renewable Groundwater Stress with GRACE." *Water Resources Research*, 51(7): 5217–5238.

Roser-Renouf, Connie, Edward Maibach, Anthony Leiserowitz, Goeff Feinberg, Seth Rosenthal, and Jennifer Kreslake. 2014. "Global Warming's Six Americas, October, 2014: Perception of the Health Consequences of Global Warming and Update on Key Beliefs." New Haven, Connecticut: Yale University and George Mason University, Yale Project on Climate Change Communication. http://environment.yale.edu/climate-communication-OFF/files/Six-Americas-October-7. Accessed on May 29 2017.

Stocker, T. F., D. Qin, G.-K. Plattner, M. Tignor, S. K. Allen, J. Boschung, A. Nauels, Y. Xia, V. Bex and P. M. Midgley. 2013. IPCC, "Climate Change 2013: The Physical Science Basis". *Contribution of Working Group I to the Fifth Assessment Report of the Intergovernmental Panel on Climate Change*. Edited by T. F. Stocker, D. Qin, G.-K. Plattner, M. Tignor, S. K. Allen, J. Boschung, A. Nauels, Y. Xia, V. Bex and P. M. Midgley. Cambridge, United Kingdom and New York: Cambridge University Press, 1535pp. Accessed September 12 2015. https://books.google.com/books?id=jn4mCAAAQBAJ&dq=IPCC+Assessment+Report+Physical+basis+2013&lr=.

UNESCO. 2016. "UN World Water Development Report 2016: Water and Jobs." World Water Assessment Programme. 2016. "UN World Water Development Report 2016: Water and Jobs." http://unesdoc.unesco.org/images/0024/002439/243938e.pdf. Accessed September 22 2017.

United Nations World Health Organization. 2017a. "Drinking-water Fact Sheet." http://www.who.int/mediacentre/factsheets/fs391/en/. Accessed on July 17 2017.

United Nations World Health Organization. 2017b. "Sanitation Fact Sheet." http://www.who.int/mediacentre/factsheets/fs391/en/. Accessed on July 17 2017.

US Department of Energy. 2013. "US Energy Sector Vulnerabilities to Climate Change and Extreme Weather." https://energy.gov/downloads/us-energy-sector-vulnerabilities-climate-change-and-extreme-weather. Accessed September 12 2017.

US Department of Energy. 2014. "Energy-Water Nexus: Challenges and Opportunities." https://www.energy.gov/under-secretary-science-and-energy/downloads/water-energy-nexus-challenges-and-opportunities. Accessed September 22 2017.

USGS. 2017. "Where Is Earth's Water?" https://water.usgs.gov/edu/earthwherewater.html Accessed August 13 2017.

Vörösmarty Charles J., P. B. McIntyre, Mark O. Gessner, David, Dudgeon, A. Prusevich, P. Green, S. Slidden, Stuart E. Bunn, Caroline A. Sullivan, C. Reidy Liermann, and P. M. Davies. 2010. "Global Threats to Water Security." *Nature* 467: 555–561.

2

Human Dimensions of Water Security

2.1 Introduction

While climate change presents formidable challenges to global water systems, water problems are primarily the result of the failure of societal institutions to manage the resource and meet the needs of current residents, the economy and environment, and future generations. Single-minded focus on climate change and hydrological response dominates water science at the expense of research that investigates how to safeguard water systems in the face of inevitable environmental and societal change. The latter emphasizes the role of markets, urban planning, supply chains, insurance schemes, policy, technology, governance, cultural attitudes, values, behaviors, institutions, legal frameworks, and decision-making strategies in mitigating water stress. It highlights the need for action to alleviate water problems, irrespective of climate change impacts. This chapter focuses on the issue of vulnerability in the water sector: what it means, why it exists, and how to remedy it.

The field of natural hazard research investigates connections between the administration of society and extreme natural events. In 1945, geographer Gilbert F. White famously declared: "Floods are 'acts of God,' but

P. Gober, *Building Resilience for Uncertain Water Futures*,
https://doi.org/10.1007/978-3-319-71234-5_2

flood losses are largely acts of man," acknowledging that humans interact with the environment to mitigate or enhance the impacts of extreme hydrological events (White 1945 p. 2). Today's vulnerabilities in water systems result from interactions among changing climate and hydrological processes, evolving human needs, and the human capacity to adapt to them. Human actions can increase flood hazard (frequency of flood events) as well as flood risk (damage from flood events). Urbanization increases runoff and this needs to be managed, as does agricultural drainage. River flood defenses in the Rhine Basin have disconnected floodplain storage from the river, leading to downstream flood hazard and programs to reconnect the river with its floodplain when possible. The impact of flood events in the river Rhine increases as climate change boosts the magnitude and frequency of flood events and as a growing number of people live in areas with high exposure to flooding (te Linde et al. 2011). Hurricane Harvey in Houston (2017) raises the question about how far we can push nature without having nature push back. Previous efforts to control storm water in Houston led to the channelization of the bayous, wetlands that protected the residential and business areas from mid-level storms. These efforts crippled the city's ability to manage a big storm like Hurricane Harvey when waters exceeded their banks and flooded residential areas, roads, and businesses. The drainage of wetlands for urban development removed the natural sponges that protected land from large flood events such as Harvey (Hernandez and Fausset 2017). Many modern developments lack sufficient open land and retention basins to absorb water from big storm events. India's financial capital, Mumbai, also experienced catastrophic storm damage in 2017 (Dhillon 2017). This cluster of these events around the world begs the question of whether flood losses are inevitable acts of god, as often depicted in the media, or the result of poor planning, lack of regulation, overbuilding in low-lying areas, and failure to anticipate the effects of a changing climate.

2.2 Vulnerability Assessments

Terms like "water stress," "water security," "water risks," and "water scarcity" are fraught with definitional ambiguity and contextual debates. The United Nation's Millennium Development Goals emphasize basic human needs, including access to an improved water source and access to improved sanitation (United Nations 2015). Cook and Bakker (2012) reviewed the emerging academic and policy literature and identified four overlapping water security themes: (1) water quality and quantity, (2) hazards and vulnerability, (3) affordability and access, and (4) sustainable development. Emphasis on water quality and quantity highlights scientific assessments of global shortage and biodiversity loss (Falkenmark et al. 2007; Vörösmarty et al. 2010). The hazards and vulnerability perspective appears in UNESCO-IHE's (Institute for Water Education) definition; it emphasizes protecting water systems from floods and droughts and safeguarding water functions and services for humans and the environment (Schultz and Uhlenbrook 2007). Affordability and access underscore the role of inequality and rising global food demands (Rockström et al. 2004; Forouzani and Karami 2011). The sustainable development definition of water security is the most holistic and includes "the availability of and acceptable quantity and quality of water for health, livelihoods, ecosystems and production, coupled with an acceptable level of water-related risk to people, environments, and economies" (Grey and Sadoff 2007, p. 548).

Policy debates about water security tolerate definitional haziness in part to allow discussion to unfold. In a pair of influential books, political scientists Deborah Stone (2002) and Frank Fisher (2003) argued that the policy process is in fact a political contest about core human values and the meaning of basic goals, including equity, efficiency, liberty, and security. Stone claimed that these goals are continuously reconstructed and contested; their varied interpretation is the object of political struggle. Ambiguous definitions provide a space that allows conflicting actors to find ways to live with their differences. It is possible for people to benefit from the same policy for altogether different reasons. Fuzzy terms such as "water security," "water stress," and "sustainable development" provide opportunities for people with diverse values, motivations, and expectations to reach consensus and move forward on collective action. Some cringe at the lack of definitional clarity, but it is sometimes part of the process of bringing people together and finding a path to reduce water system vulnerabilities.

Geoscientist Charles Vörösmarty and colleagues produced two highly influential vulnerability assessments of the state and future of global water resources (Vörösmarty et al. 2000, 2010). In the first, the team estimated that one-third of the world's population now lives in a state of high water stress (here defined as a situation in which water use or demand constitutes more than 40% of river discharge or supply). Using various scenarios of climate change and assumptions about population and economic growth and the intensification of water use into the future, they concluded that changes in population and economic development over the next 25 years (1995–2025) are far more important than potential changes in the mean climate in affecting the future balance of water demand and supply.

In subsequent research involving both water security and river biodiversity, they found that in 2000 nearly 80% of the world's population lived in areas with high risk of either human water security threat or biodiversity threat (Vörösmarty et al. 2010). Regions of intensive agriculture and dense settlement show high incident threat, as experienced in the US, Europe, large portions of Central Asia, the Middle East, Indian subcontinent, and eastern China. Very few of the world's river basins are unaffected by humans and their activities and thus impervious to threats from pollution, catchment disruption, water resource problems, or biotic factors. They note that highly developed countries like the US and western European nations have reduced risk through massive investments in water infrastructure. The story is different, however, in developing countries where the risk of water threats remains high, especially in Africa, Central Asia, China, India, Peru, and Bolivia. Lack of water security is manifest in problems of unsafe drinking water and inadequate sanitation but also in problems of food shortage, slow economic growth, and energy insecurity. Failure to invest in infrastructure and inability to protect environmental flows threatens both human water security and biodiversity.

Water resource shortages affect people directly via access to drinking water and sanitation services but also indirectly through the industries in which they work. Eight water and natural resource-dependent industries, agriculture, forestry, fisheries, energy, resource-intensive manufacturing, recycling, building, and transport, account for half of the global workforce. More than a billion people work in inland fisheries and aquaculture, agriculture, and forestry. The latter two are most threatened by freshwater disruptions (World Water Development Report 2016, p. 10).

The potential for water crisis emerged as one of the World Economic Forum's most likely and impactful global risks, along with issues such as interstate conflict, cyber warfare, the spread of infectious diseases, weapons of mass destruction, and failure of climate change adaptation. What emerges from its most recent reports is the interconnectedness of local water problems to other looming risks that create far-flung impacts for the economy and society. In July 2011, Thailand experienced a major flood event resulting in $46.5 billion of economic damage (World Bank 2012). Heavy rains in early summer filled reservoirs rendering them unable to cope with the onset of monsoon storms in late July. Above-normal rainfall exacerbated drainage problems throughout the fall, and channels of the Chao Phraya River were unable to cope with high flows. Significant downstream flooding occurred, particularly in the Bangkok area. Bangkok itself developed on floodplains where new industrial parks replaced natural waterways (Haraguchi and Lall 2015). The flood event had significant impacts for Thailand, resulting in an estimated 815 deaths, 2.5 million displaced people, 19,000 destroyed homes, and 17.578 square kilometers of impacted farmland (World Bank 2012; Government of Thailand 2011).

Consequences also occurred outside of Thailand as the water crisis rippled through international supply chains for automobile parts and electronic equipment. Thailand has become a major manufacturing center due to government incentives, tax breaks, and land acquisition deals specifically designed to lure automotive companies and high-tech manufacturers. It produces the parts that go into making cars and electronic equipment, and thus the economic effects of the flooding spread globally. Japanese automobile manufacturers faced shortages in key parts produced in Thailand; Toyota and Honda curtailed production at automobile plants in North America because of the inability to secure Thai parts; and computer companies such as Lenovo and Samsung were unable to obtain the hard disk drives that go into the making of their computers. Global supply chain problems in these sectors persisted into early 2012. The widespread impacts of the Thai flooding for global supply chains raised awareness in the private sector of the potential for localized water problems to cause global economic disruptions. The World Bank estimated that Thai flooding lowered the global gross national product (GNP) growth rate in 2011 from 4.1% to 2.9% (World Bank 2012).

Globalization and the drive for improved efficiencies led businesses to reduce inventory, shorten transportation lines, and rely on a small number of suppliers. In the drive to be more efficient, the supply chains became more brittle, and thus vulnerable to a localized water crisis such as Thai flooding. Haraguchi and Lall (2015) found differences across economic sectors (automotive parts versus hard drive disk production) and individual producers (Toyota versus Honda and Nissan) that point to management strategies, such as alternative procurement mechanisms, multiple suppliers, diverse transportation connections, and collaboration between producers and suppliers, that limited production time lost to flood disruptions. Thai floods were a wake-up call for the private sector that local water vulnerabilities can translate into global disruptions with significant economic impacts.

In October 2016, seven international food companies announced plans to reduce water use and pollution impacts in their supply chains (AgriPulse 2016). They are part of a collaboration organized by the World Wildlife Fund and Ceres, a non-profit corporation in Boston, Massachusetts, aiming to build capacity and leadership in corporations to address sustainability challenges. One-third of the world's food grows in areas of high water stress or competition, and agriculture is a major cause of pollution worldwide. As part of this new initiative:

1. PepsiCo will work with its agricultural suppliers to improve the water-use efficiency of its direct agricultural supply chain by 15% by 2025 (compared to 2015) in high-water-risk sourcing areas, including India and Mexico.
2. Hain Celestial, an American company specializing in food production and personal care products, will strengthen water and fertilizer management practices of farmers in its supply chains.
3. Hormel Foods will develop a comprehensive water stewardship policy, setting water management expectations that go beyond regulatory compliance for its major suppliers, contract animal growers, and feed suppliers.
4. WhiteWave Foods will develop a road map for agricultural water stewardship over key commodities (dairy, soy, almond, and produce) in areas of high water risk, including California.

5. Diageo, General Mills, and Kellogg will conduct water risk assessments for reducing water risk in agricultural supply chains, set reduction goals, and support producers in addressing these issues.

2.3 Natural Hazards and Vulnerability

Water problems often come to public attention during extreme natural events, such as the Thai floods. Natural hazards, by definition, exceed the human capacity to cope and disrupt the normal functioning of society and economy. Sudden onset hazards appear rapidly and last for a period ranging from hours to weeks. They include floods, earthquakes, tornados, and wildfires. Chronic hazards are slow in onset and are barely noticeable to society; they include drought, sea level rise, and weather events that may or may not be associated with climate change. Chronic hazards are often harder to interpret because they are difficult to separate from long-term, structural changes in society, economy, and environment, such as urbanization, deforestation, and globalization. They affect populations slowly, and become disasters only after they reach a tipping point (Cutter et al. 2009). Water-related hazards include floods, droughts, water quality episodes, and sea level rise.

The early study of hazards focused on their biophysical aspects, return periods, and emergency management. Even today, the focus of press attention is often on the anatomy of a particular storm before and during the flood event when coverage is most intense. In the 1970s and 1980s, political ecologists attempted to de-emphasize the extreme natural event itself and focus on the societal conditions that put people in unsafe places, limit support for them, and undermine their ability to cope. The concept of vulnerability frames this new discussion. Vulnerability is, by definition, the diminished capacity to cope with extreme natural events. Vulnerability assessments include three dimensions: (1) Physical exposure is the degree to which a natural event can harm people and property. This is what natural scientists focus on. (2) Sensitivity involves social characteristics such as poverty status, low income, age, and health status that limit the ability to cope with an extreme event. Recent Hurricane Irma in Florida has drawn attention to reasons that people do not evacuate dangerous areas in the face of an oncoming hurricane. They lack financial assets, are in poor health, care for someone who is in poor health, or lack the social network needed to coordinate a last-minute 400-mile ride. (3) Adaptive capacity includes emergency management, community engagement, strong public institutions, and ability to translate past experiences into social action (Adger et al. 2005; Polsky et al. 2007). Social scientists emphasize sensitivity and adaptive capacity with physical exposure as a backdrop to the larger human drama of engaging extreme natural events.

Context is important in how hazards are experienced and vulnerabilities are reduced (Wisner et al. 2004). People in developing countries, for example, are at significant risk of loss of life and livelihood from tropical cyclones. In more developed countries, there is significant risk of property and infrastructure damage, although 1,833 people died in New Orleans in the 2005 Hurricane Katrina event (Zimmerman 2015). The hazard field relies heavily on case studies and narratives of particular events to form conclusions and develop frameworks for vulnerability assessments.

The Pressure and Release Model represents hazard vulnerability as an evolving set of interrelated environmental and social processes (Wisner et al. 2004). Vulnerability begins with root causes (e.g., power relations, economic systems, political systems). Dynamic pressures such as rapid population growth, increasing urbanization, and neo-liberal reforms (e.g., laissez-faire markets, unrestricted trade, cuts in government spending, privatization) translate these pressures into unsafe conditions. Unsafe conditions include people living in places without safe drinking water or sanitation; people living on hillslopes, in floodplains, and in coastal zones unable to afford insurance or easily evacuate their homes in an emergency; and people living under governments that are unwilling or unable to play viable roles in hazard mitigation and emergency management. Disaster occurs when this chain of root causes, dynamic pressures, and unsafe conditions intersects with extreme natural events. Thus, vulnerability evolves slowly over time through human action and inaction only to transform into disaster by some climatic or geologic event.

Vulnerability often has a social justice dimension. Mexico City, for example, strains to provide reliable water services to its population. Aztecs built the city on a network of lakes. They expanded it over time with landfill and planted crops on floating gardens called *chinampas*. The Spaniards drained the lakes, and replaced the dykes and canals with streets and squares. Today, the city is a jumble of neighborhoods, and there is no centralized system of water provision and waste disposal. Some 20% of residents lack access to a reliable supply of drinking water and must hire trucks to deliver water at costs much higher than what wealthier residents in better-served neighborhoods pay for their water (Kimmelman 2017). Hurricane Harvey also showed the logistical challenges facing people without cars, family, and the financial means to evacuate in the face of extreme weather events.

Social and environmental scientists have begun to anticipate damage from hazardous events based on analysis of vulnerabilities. Di Baldassarre et al. (2015) used scenarios (stories of possible futures) to compare flood risk in technological societies that emphasize building dams and levees to protect people and property with green societies that resettle people and economic activity out of flood-prone areas. They used the concept of social memory—the ability of humans to process and recall the deleterious effects of flooding—to explain the differing paths of green versus technological societies and simulate how flood damage occurs over time. They found that technological societies can protect themselves from low-impact events, but green societies are less vulnerable to high-impact events. Over the long term, investment in flood infrastructure reduces the perception of risk and lowers social memory, thereby increasing vulnerability to high-impact events. Hazard researchers call this the "levee effect" whereby short-term investment in flood mitigation infrastructure encourages more people to resettle close to the river to gain economic advantage, but in the long term increases the amount of damage associated with large events (Kates et al. 2006; Burton and Cutter 2008; Montz and Tobin 2008).

The following case studies of hazard events in California (drought), Calgary in western Canada (flooding), and the Aral Sea in Central Asia (overuse and environmental collapse) illustrate the powerful hand of human agency in exposing people and property to water hazards. In each case, societal institutions failed to anticipate the limits to growth with significant and unequal impacts on local and regional populations.

2.4 California Drought 2011–2015

Droughts are a common feature of California's climate. The drought episode between the fall of 2011 and early 2016 was the driest since record keeping began in 1895; 2014 and 2015 were the two hottest years in the state's history (Hanak et al. 2016). California governor Edmund Gerald Brown Jr. declared statewide drought emergencies in January 2014 and in April 2015 and ordered a 25% reduction (relative to 2013 use) in water use for cities and towns. Urban communities were relatively well prepared, with sharing agreements in place and having made investments in infrastructure. Per capita water use declined sharply in urban areas during the drought, and most cities were able to cope. Growers in rural areas received 50% less irrigation water than usual in 2015 and pumped groundwater to compensate for reduced surface supplies with significant negative consequences for wells in rural communities and the environment. Extinction threatens 18 species; wildlife refuges experienced shortfalls, and wildfires endangered dry, dense forests. Seen from one angle, damage from the recent drought was the result of an unusual set of climatic circumstances. From another angle, it was a disaster waiting to happen as the state grew rapidly, putting ever more pressure on its water resources without adequately adjusting allocation schemes and policies to accommodate growth, rapid urbanization, and increasing vulnerability to drought.

Drought is a recurrent feature of California's climate (Fig. 2.1). The amount of geographic area that was "abnormally dry" (a condition characteristic of coming into or out of drought) varied after 2000, with a large portion of the state experiencing drought after 2007. The amount of land exposed to extreme or exceptional drought was more limited during the first two episodes but quite substantial between 2014 and 2016. Extreme drought translates into major losses for crops and pastures and widespread water shortages and restrictions. Exceptional drought adds water shortages

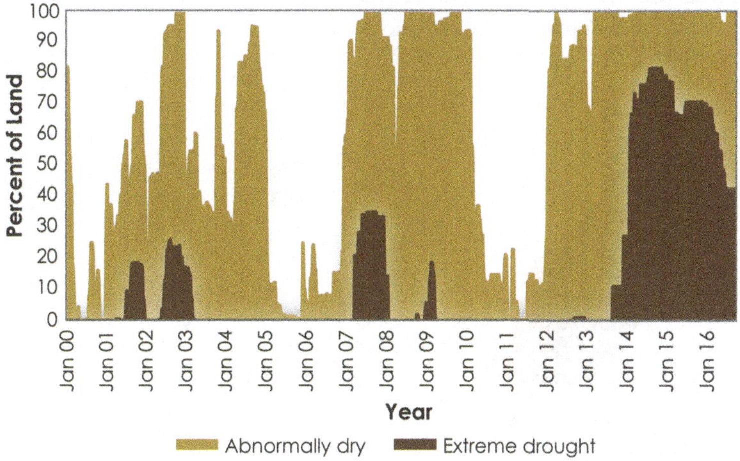

Fig. 2.1 Percentage of California's land area affected by abnormally dry and extreme drought conditions, 2000–2016. United States Drought Monitor/http://droughtmonitor.unl.edu/

in reservoirs, streams, and wells, causing water emergencies, and has a Palmer Drought Severity Index (PDSI) below –5. PDSI measures dryness based on temperature and precipitation. Zero is normal; 2 is moderate drought; 3 is severe drought; 4 is extreme drought; and 5 is exceptional drought (National Oceanographic and Atmospheric Administration 2017). The recent drought was particularly harsh and widespread in a state of almost 40 million residents (Fig. 2.2a, b). During most of 2015, more than two-thirds of California's land area was in extreme to exceptional drought.

While it is easy and timely to focus on the climatic and geographic aspects of drought and their immediate impacts on society, the economy, and the environment, it is also important to see the recent water crisis in California in a historical context. Late nineteenth- and early twentieth-century water infrastructure development (dams, reservoirs, and canals) enabled irrigated agriculture in California and western North America. Early nineteenth-century explorer and head of the Geologic Survey John Wesley Powell saw the key role of water in shaping the destiny of the American West and recommended that development of this arid region be limited to the relatively small land areas irrigated by river flows. In a report to Congress published in 1876, Powell urged

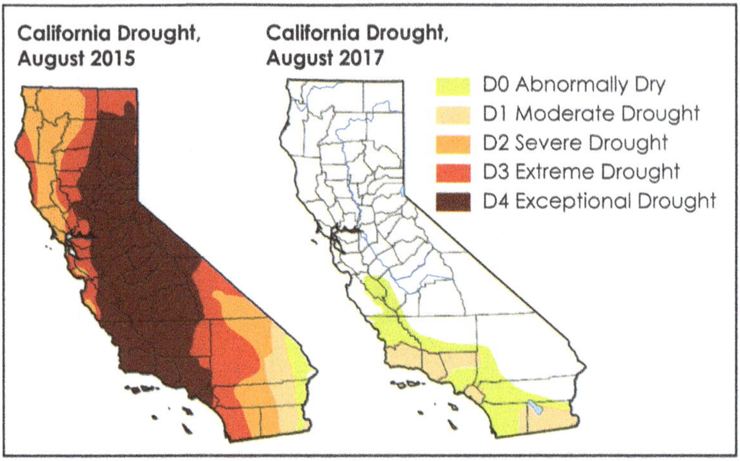

Fig. 2.2 California drought conditions in (**a**) August 2015 and (**b**) August 2017. United States Drought Monitor/http://droughtmonitor.unl.edu/

that development occur in only small areas supported by river irrigation. National leaders in Washington ignored Powell's vision of western water and land development, and, in the rush to settle the country, constructed large-scale, federally funded dams, reservoirs, and canals. It was US federal policy to construct water projects on virtually every flowing river in the western states (Reisner 1986). These projects initially supported irrigated agriculture in downstream valleys and later large-scale urbanization of the West. It would be unthinkable to find cities like Los Angeles (13.3 million), Las Vegas (2.1 million), and Phoenix (4.5 million) today in this arid and semi-arid region without the large-scale water infrastructure that was built to support early agricultural development (US Census 2017).

A first-in-time-first-in-right (FITFIR) allocation system has governed water use in California and the rest of western North America since European settlement. It assigns the highest priority to users who first put water to "beneficial" use, beneficial defined in terms of value to humans. This system assured early settlers of water supplies to entice their investment in the risky venture of settling a new country. Today, farmers with more recent or junior rights have lower priority in times of shortage.

In 2014, California's State Water Resources Control Board, which administers water rights and quality standards, curtailed water diversions by many junior water-rights holders for the first time since 1977. These orders extended to senior rights holders in 2015.

Also significant from a policy and regulatory perspective was that cities and farmers are allowed to augment supplies during drought years by pumping groundwater. In a typical year, groundwater supplies account for about one-third of total farm and urban water. After 2014, this share exceeded 50%. Extra pumping exacerbated chronic groundwater overdraft in the highly productive agricultural region of the Central Valley, leading to falling water tables, increasing pumping costs, and drying up of domestic wells (Hanak et al. 2016). The state passed the Sustainable Groundwater Management Act in 2014 to empower local authorities to develop sustainable groundwater management plans by 2020 and implement them by 2042 (Water Education Foundation 2015), but the impacts of these policies were too late to affect 2011–2016 water resource issues.

Viewed at this longer time scale, California's recent drought damage is less an unusual climatic phenomenon and more an inevitable outcome of perpetual growth, excessive use, business-as-usual management, and inadequate drought preparation. In his iconic 1986 story of western water development in *Cadillac Desert*, journalist Marc Reisner noted that early water infrastructure expansion in American West set in motion a pattern of development that reduced the short-term unreliability of western water systems but increased vulnerabilities to water shortage in the long term. This form of development "though amazingly fruitful in the short run, leaves everyone and everything more vulnerable in the end" (Reisner 1986, p. 499).

Past adaptation to California droughts (1976–1977, 1988–1992, and 2007–2009) led to active conservation programs, greater use of water markets, increased groundwater extraction, more intensive irrigation, more infrastructure for conveyance and storage, and more reuse and desalination (AghaKouchak et al. 2015). These efforts wrung redundancies out of the water system in an effort to improve efficiencies, rendering it highly vulnerable to climate shocks and future growth pressure (California's population grew from 27.1 million in 1986 to 39.1 million in 2015). In the absence of groundwater regulation, excessive extraction during the recent drought lowered the base flows of rivers and streams, disconnected flow networks, and compromised the habitats of some native fish species.

Policy discussion in light of the recent drought in California includes awareness of the need to prepare more effectively for future droughts and manage groundwater sustainably. In a 2016 Executive Order, Governor Brown declared that water conservation efforts put into place during the recent drought would remain permanent and the state will adjust to the "new normal." "Ongoing drought conditions and our changing climate require California to move beyond temporary emergency drought measures and adopt permanent changes to use water more wisely and to prepare for more frequent and persistent periods of limited water supply" (Executive Department State of California 2016). The policy discussion also includes the very difficult decisions ahead involving human water uses versus water to support fish and wildlife and the value of irrigated agriculture that consumes 80% of California's water but accounts for only 2% of its economic production (Mount and Hanak 2016). Other significant issues include value added by investing in more diversified supplies, capacity of water markets to facilitate high-value water use, and managing groundwater as a sustainable resource. Protecting wells in poor rural communities from excessive groundwater use and balancing the needs of farmers to trade water from the San Joaquin delta with responsibility to protect endangered species of native fish are also on the policy agenda (Public Policy Institute of California 2017). Tackling these issues moves the water policy agenda from the need for better climate predictions to preparing water management for more severe, persistent, and uncertain water futures.

2.5 Calgary Flood 2013

Flooding is the counterpoint to drought in many semi-arid regions of the world, including in the Canadian city of Calgary at the confluence of the Bow and Elbow Rivers in the province of Alberta (Fig. 2.3). Calgary has a long history of major flooding (1883, 1884, 1897, 1902, 1915, 1923,

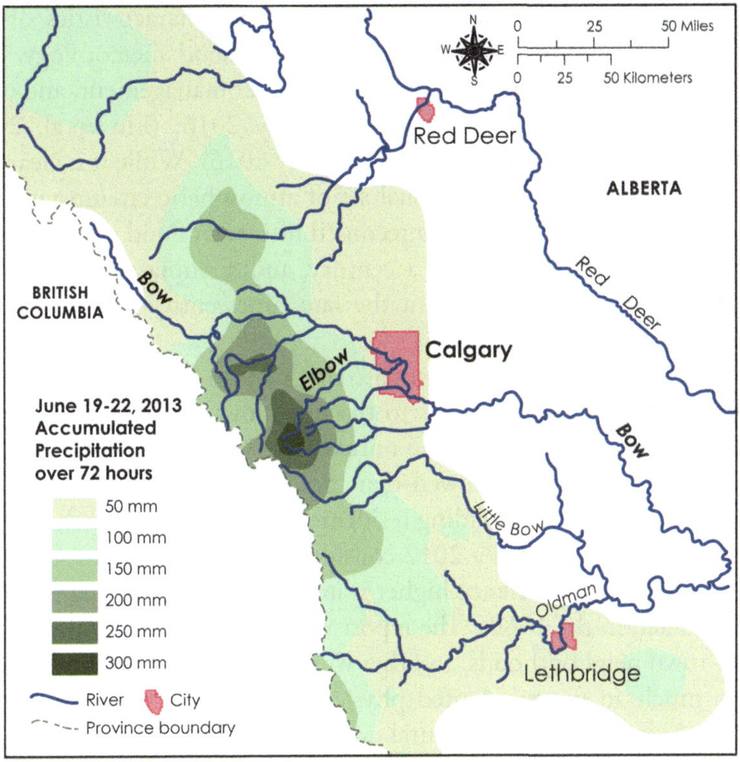

Fig. 2.3 Calgary is located at the confluence of the Elbow and Bow Rivers downstream from the Canadian Rocky Mountains

1929, 1932, 1950, 2005, 2013), especially between late May and early July when snow melt and rainfall runoff are discharged from the nearby foothills of the Canadian Rocky Mountains and feed the rivers that flow through the city's center. A large storm hit Calgary in June 2013 causing catastrophic flooding (Liu et al. 2016), claiming the lives of four people and the evacuation of 100,000. It was the costliest natural disaster in Canadian history (until the Fort McMurray wildfire in 2016 and floods in eastern Canada in 2017), resulting in an estimated cost of $6 billion CAD (CBC News 2014). The downtown business center was particularly hard hit, with the storm disrupting the regional transportation system, closing businesses, and interrupting everyday life in this regional capital of 1.4 million people in 2015 (Statistics Canada 2016).

Climatologists and hydrologists dissected the characteristics of this storm, including atmospheric conditions and flood meteorology, land surface processes associated with the flood, water management, and operational decision-making (Fang and Pomeroy 2016; (Liu et al. 2016; Pomeroy et al. 2016; Whitfield and Pomeroy 2016). While it is clear that the flood resulted from an unusual set of atmospheric circumstances, it was not the largest such event in recorded history. A flood of comparable magnitude occurs about twice a century, and a similar flood occurred previously on the Bow River in the late nineteenth century (City of Calgary 2014).

Downtown Calgary was inundated to depths never before seen in modern history, power supplies to the downtown and some residential neighborhoods stopped, and the normal activities of city life came to a halt. Post-event analysis showed that the city's flood-hazard maps were woefully inaccurate in providing residents an accurate assessment of their flood risk. It also revealed a 2012 consultant report that had warned the city that it would experience higher water levels and worse damage than was predicted in the 1980s. The report warned of the inundation of several in-town neighborhoods. In other words, damage from this event was not so much an unprecedented physical event but a failure to prepare adequately for a predicable natural event (CBC News 2013).

Calgary grew rapidly after 1980, owing to a natural resource boom in western Canada (Fig. 2.4). As a result, many of its 2013 residents had relatively little social memory of or experience with severe flood conditions, even including a 2005 event that caused damage to around 40,000 homes, resulted in evacuations of 1500 households, and washed away bridges and attached infrastructure. Many residents were newcomers with no real sense of the city's flood history.

City leaders reflected on how to organize flood response moving forward and appointed an independent expert panel to undertake a flood mitigation investigation and recommend future action. The report concluded that, although there will always be risk of damage associated with a city that occupies a floodplain, the current level of risk was "unacceptably high," and it was time to invest in better risk management and reduce flood vulnerabilities (City of Calgary, Calgary's Flood Resilient Future 2014). The report acknowledged that risk mitigation involves

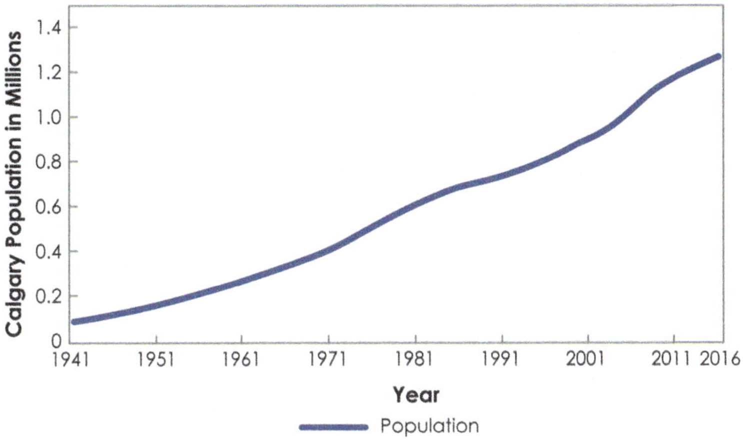

Fig. 2.4 Rapid post-war population growth in Calgary. Statistics Canada (2016)

hard choices that might affect some people negatively. "Achieving higher flood protection… could require greater investment in flood barriers, stricter land-use planning, additional requirements and limitations for development in flood risk areas and large capital works, among other actions. Many of these measures are expensive and disruptive and some would also have aesthetic and environmental impacts" (City of Calgary 2014, p. 19). This discussion exemplifies the role of public engagement when deciding what level of risk is acceptable and how to distribute risks and costs across society. These are not engineering problems alone; they are social justice, aesthetic, and public policy questions that can only be resolved through public discussion and collective action.

The city adopted efforts to reduce potential flood damage by updating flood hazard risk maps and ensuring that updated maps are publically accessible, revising land-use planning policies that limit floodplain development, removing buildings from the floodplain, and developing computer models of groundwater movement. The effort also improved forecasting and warning systems, expanded observation networks and flood alerts, improved communications with the public, increased the capacity of an upstream reservoir, and increased investments in a large-scale capital works project that would divert water from the downtown areas.

This case also speaks to deep societal divides that underlie who bears the risks and costs associated with natural hazards. Overland flood insurance has not been available to Canadian homeowners until recently (Toronto Sun 2017). Governments have created financial assistance programs to help Canadian homeowners after flood events, but coverage is helter-skelter, and homeowners lack clarity on coverage. Most developed countries have privately funded flood insurance because risk-based premiums and deductibles encourage homeowners to take action to reduce floods risk. Lacking national markets and schemes for flood insurance, individual Canadians had been expected to assume more responsibility for flooding than is the case in countries with more organized, national-level flood insurance programs (Sandink et al. 2010). Calgary expects individual homeowners to assess floodplain maps and understand their risk of damage from flood events. These maps are available via the city's Website.

2.6 Aral Sea

Environmental collapse of the Aral Sea in Kazakhstan and Uzbekistan is arguably the world's most preventable natural disaster (Fig. 2.5). The Aral Sea was in 1969 the world's fourth largest lake in area, comprising 68,000 km^2 with an average depth of 16 m (Micklin 1988). It supported bustling fishing villages, and in its heyday produced one-sixth of the former Soviet Union's fresh fish and employed some 40,000 workers. In a five-year plan to support irrigated agriculture in this desert region, the former Soviet Union diverted water from the Aral Sea's two major rivers, the Amu Darya and Syr Darya, to grow cotton. In 1988, Uzbekistan was the world's largest exporter of cotton. By 1960, farmers used between 20 and 60 km^3 of water each year to irrigate crops, and much of the water evaporated instead of returning to the sea. The Aral Sea began to shrink. From 1961 to 1970, the Aral's level fell at an average of 20 cm (7.9 in) a year. In the 1970s, the average rate of decline nearly tripled to 50–60 cm (20–24 in) per year, and by the 1980s, it continued to drop with a mean of 80–90 cm (31–35 in) each year. By 2000, the lake itself was a sliver of its former size, and even that area is gradually disappearing (Fig. 2.5).

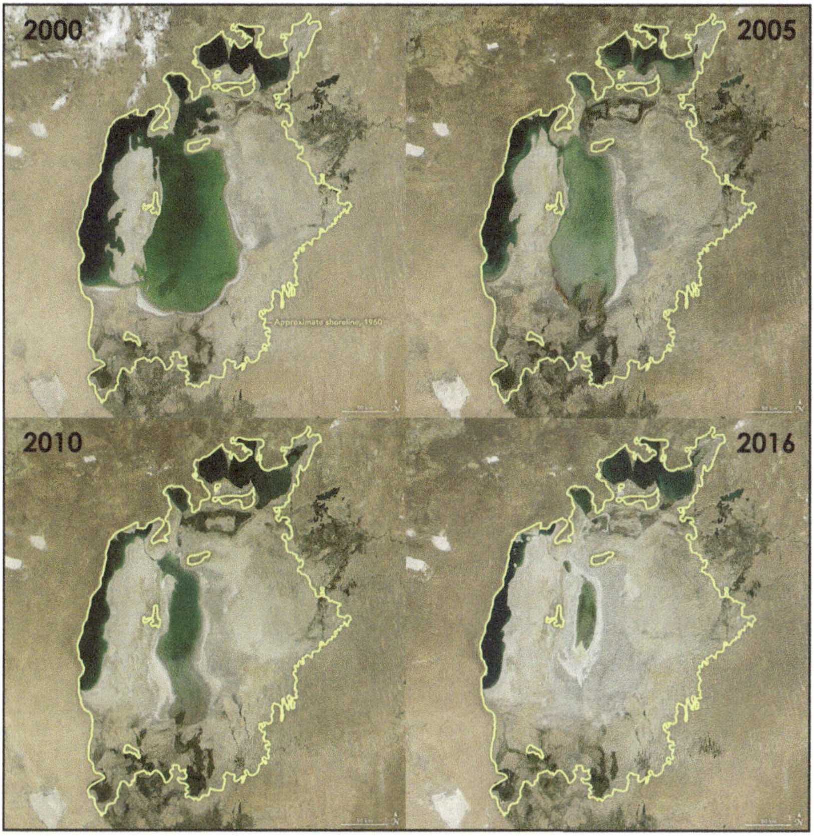

Fig. 2.5 Aral Sea, 2000–2016. NASA/https://earthobservatory.nasa.gov/Features/WorldOfChange/aral_sea.php

With overuse and diversion, levels of salinity rose dramatically. The dry lake's bed covered in salt gave rise to problems of blowing salt and dust storms, loss of biological productivity, and a halt in commercial fishing due to toxicity from chemicals and fertilizers. Former fishing villages are now miles from the shore, and most are abandoned. The iconic symbol of Aral Sea desiccation is boats beached on salt flats and harbors now many miles from water. Health consequences are dire, with upticks in tuberculosis resulting from blowing dust and liver, kidney, and eye problems attributed to the dust storms.

Efforts to revive the sea are under way, but the damage has been severe and irreversible. Independent countries, Kazakhstan, Uzbekistan, Turkmenistan, Tajikistan, and Kyrgyzstan signed a deal in 1979 to pledge 1% of their budgets to help the sea recover. In March 2000, UNESCO presented their "Water-related vision for the Aral Sea basin for the year 2025" at the second World Water Forum in The Hague.

2.7 Conclusions

Extreme natural events and increasing human interventions pressured water systems in California, Canada, and Central Asia. The three case studies just described were preventable and predictable. They resulted more from the human failure to anticipate the impacts of ongoing development than from any particular biophysical event. They share patterns of overuse and rapid growth creating vulnerability to natural forces. They illustrate the value of taking precautionary action to reduce vulnerabilities in the face of climatic and other forms of uncertainty. Chapter 3 lays out the special problem of climatic uncertainty for water science and policy.

References

Adger, W. Neil, Terry P. Hughes, Carl Folke, Stephen R. Carpenter, and Johan Rockström. 2005. "Social-Ecological Resilience to Coastal Disasters," *Science* 309(5737): 1036–1039.

AghaKouchak, Amir, David Feldman, Martin Hoerling, and Travis Huxman. 2015. "Recognize Anthropogenic Drought." *Nature* 524(7566): 409–411.

Agripulse. 2016. "Food Companies Target Water Risks in global supply chains." http://www.foodandagpolicy.org/news/story/food-companies-target-water-risks-global-supply-chains-agri-pulse. Accessed September 22 2017.

Burton, Christopher, and Susan L. Cutter. 2008. "Levee Failures and Social Vulnerability in the Sacramento-San Joaquin Delta Area, California." *National Hazards Review* 9(3): 136–139.

CBC News. 2013. "2012 Calgary Flood Study Foresaw June's Devastation." December 10, 2013. http://www.cbc.ca/news/canada/calgary/2012-calgary-flood-study-foresaw-june-s-devastation-1.2458014. Accessed September 22 2017.

CBC News. 2014. "Alberta flood: By the Numbers One Year Later." June 15, 2014. http://www.cbc.ca/news/canada/calgary/alberta-flood-by-the-numbers-one-year-later-1.2676657. Accessed September 22 2017.

City of Calgary. 2014. "Calgary's Flood Resilient Future: Report from the Expert Management Panel on River Flood Mitigation June 2014." http://www.calgary.ca/_layouts/cocis/DirectDownload.aspx?target=http%3a%2f%2fwww.calgary.ca%2fUEP%2fWater%2fDocuments%2fWater-Documents%2fFlood-Info-Documents%2fExpert-Management-Panel-Report-to-Council.pdf&noredirect=1&sf=1. Accessed September 22 2017.

Cook, Christina, and Karen Bakker. 2012. "Water Security: Debating an Emerging Paradigm," *Global Environmental Change* 22: 94–102.

Cutter, Susan L., Christopher T. Emrich, Jennifer J. Webb, and Daniel Morath. "Social Vulnerability to Climate Variability Hazards: A Review of the Literature." *Final Report to Oxfam America* 5(2009): 1–44.

Dhillon, Amrit. 2017. "South Asian Floods: Mumbai Building Collapses as Monsoon Rains Wreak Havoc." *The Guardian*, August 31 2017. https://www.theguardian.com/world/2017/aug/31/south-asia-floods-fears-death-toll-rise-india-pakistan-mumbai-building-collapses. Accessed September 27 2017.

Di Baldassarre, Guiliano, Alberto Viglione, Gemma Carr, Linda Kuil, Kun Yan, Luguia Brandimarte, and Gunter Blöschl. 2015. "Debates—Perspectives on Socio-Hydrology: Capturing Feedbacks Between Physical and Social Processes." *Water Resources Research* 51(6): 4770–4781.

Executive Department State of California. 2016. "Executive Order B-37-16 Making Water Conservation a California Way of Life." https://www.gov.ca.gov/docs/5.9.16_Executive_Order.pdf. Accessed September 22 2017.

Falkenmark, M., A. Berntell, A.Jägerskog, J. Lundqvist, M. Matz, and H. Tropp. 2007. *On the Verge of a New Water Scarcity: A Call for Good Governance and Human Ingenuity*. Stockholm: Stockholm International Water Institute.

Fang, Xing, and John W. Pomeroy. 2016. "Impact of Antecedent Conditions on Simulations of a Flood in a Mountain Headwater Basin." *Hydrological Processes* 30(16): 2754–2772.

Fisher, Frank. 2003. *Reframing Public Policy: Discursive Politics and Deliberative Practices*. New York: Oxford University Press.

Forouzani, Masoumeh, and Eztollah Karami. 2011. "Agricultural Water Poverty Index and Sustainability." *Agronomy for Sustainable Development* 31(2): 415–431.

Government of Thailand. 2011. "24/7 Emergency Operation Center for Flood, Storms and Landslide. Current Situation," No. 96/2011, December 16, 2011. http://www.adrc.asia/documents/disaster_info/2011/12/EOC_Report_16_Dec-eng.pdf. Accessed September 22 2017.

Grey, David, and Claudia W. Sadoff. 2007. "Sink or Swim: Water Security for Growth and Development." *Water Policy* 9(2007): 545–571.

Hanak, Ellen, Jeffrey Mount, and Catrin Chapelle. 2016. "California's Latest Drought." Public Policy Institute of California. http://www.adrc.asia/documents/disaster_info/2011/12/EOC_Report_16_Dec-eng.pdf. Accessed on September 22, 2017.

Haraguchi, Masahiko, and Lall Upmanu. 2015. "Flood Risks and Impacts: A Case Study of Thailand's Floods in 2011 and Research Questions for Supply Chain Decision Making." *International Journal of Disaster Risk Reduction* 14(3): 256–272.

Hernandez, Manny, and Richard Fausset. "A Storm Forces Houston, the Limitless City, to Consider Its Limits." *New York Times*, August 30, 2017. Accessed August 31 2017.

Kimmelman, Michael. 2017. "Mexico City Parches and Sinking Faces a Water Crisis." *The New York Times*, June 29, 2017. https://www.nytimes.com/2017/08/30/us/houston-flooding-growth-regulation.html. Accessed September 12 2017.

Kates, Robert W., Craig E. Colten, Shirley Laska, and Stephen P. Leatherman. 2006. "Reconstruction of New Orleans After Hurricane Katrina: A Research Perspective," *Proceedings of the National Academy of Sciences*, 103(40): 14653–14660.

Liu, A. Q., C. Mooney, K. Szeto, J. M. Thériault, B. Kochtubajda, R. E. Stewart, S. Boodoo, R. Goodson, Y. Li, and J. Pomeroy. 2016. "The June 2013 Alberta Catastrophic Flooding Event: Part 1–Climatological Aspects and Hydrometeorological Features." *Hydrological Processes* 30(26): 4899–4916.

Micklin, Phillip P. 1988. "Dessication of the Aral Sea: A Water Management Disaster in the Soviet Union." *Science* 241(4870): 1170–1176.

Montz, Burrell E., and Graham A. Tobin. 2008. "Livin' Large with levees: Lessons learned and Lost." *Natural Hazards Review* 9(3): 150–157.

Mount, Jeffrey, and Ellen Hanak. 2016. "Water Use in California." Public Policy Institute of California. http://www.ppic.org/publication/water-use-in-california/. Accessed August 8 2017.

National Oceanic and Atmospheric Administration (NOAA). 2017. "Palmer Drought Severity Index, Explanation." http://www.cpc.ncep.noaa.gov/products/analysis_monitoring/cdus/palmer_drought/wpdanote.shtml. Accessed September 22 2017.

Polsky, Colin, Rob Neff, and Brent Yarnal. 2007. "Building Comparable Global Change Vulnerability Assessments: The Vulnerability Scoping Diagram," *Global Environmental Change*, 17(3–4): 472–485.

Pomeroy, J. W., X. Fang, and D. G. Marks. 2016. "The Cold Rain-on-Snow Event of June 2013 in the Canadian Rockies—Characteristics and Diagnosis." *Hydrological Processes* 30(17): 2899–2914.

Public Policy Institute of California. 2017. "California Faces Growing Water Management Challenges." Accessed July 18, 2017. http://www.ppic.org/content/pubs/report/R_212EHR.pdf. Accessed July 18 2017.

Reisner, Marc. 1986. *Cadillac Desert: The American West and Its Disappearing Water*. New York: Viking Press.

Rockström J., C. Folke, L. Gordon, N. Hatibu, G. Jewitt, F. Penning de Vries, F. Rwehumbiza, H. Sally, H. Savenije, and R. Schulze. 2004. "A Watershed Approach to Upgrade Rainfed Agriculture in Water Scarce Regions through Water System Innovations: An Integrated Research Initiative on Water for Food and Rural Livelihoods in Balance with Ecosystem Functions." *Physics and Chemistry of the Earth* 3(2): 1109–1118.

Sandink, Dan, Paul Kovacs, Gregg Oulahen, and Glenn McGillivray. 2010. "*Making Flood Insurable for Canadian Homeowners*." Institute for Catastropic Loss Reduction & Swiss Reinsurance Company. https://www.iclr.org/images/Making_Flood_Insurable_for_Canada.pdf. Accessed January 1 2018.

Schultz, Bart, and Stefan Uhlenbrook. 2007. "Water Security: What Does It Mean, What May It Imply? Water for a Changing World – Developing Local Knowledge and Capacity." In *Developing Local Knowledge and Capacity*, edited by Guy Alaerts and Nicolas Dickinson, 27–40. Delft, Netherlands: UNESCO Proceedings.

Statistics Canada. 2016. "Population Estimates and Projections." http://www.statcan.gc.ca/tables-tableaux/sum-som/l01/cst01/demo05a-eng.htm. Accessed June 1 2017.

Stone, Deborah. 2002. *Policy Paradox: The Art of Political Decision Making*. New York: Norton.

te Linde, A. H., P. Bubeck, J. E. C. Dekkers, H. de Moel, and J. C. J. H. Aerts. 2011. "Flood Risk Estimates Along the River Rhine." *Natural Hazards Systems Science* 11: 459–473.

Toronto Sun. 2017. "Homeowners Could be on Hook for Flooding: Insurance Experts." http://www.torontosun.com/2017/05/08/homeowners-could-be-on-hook-for-flooding-insurance-experts. Accessed August 9 2017.

United Nations. 2015. "The Millennium Development Goals Report 2015." http://www.un.org/millenniumgoals/2015_MDG_Report/pdf/MDG%20 2015%20rev%20(July%201).pdf. Accessed September 22 2017.

US Bureau of the Census. 2016. "Annual Estimates of the Resident Population: April 1 2010 to April 1, 2016." https://factfinder.census.gov/faces/tableser-

vices/jsf/pages/productview.xhtml?pid=PEP_2016_PEPANNRES&src=pt. Accessed June 22 2017.
Vörösmarty Charles J., Pamela Green, Joseph Salisbury, and Richard B. Lammers. 2000. "Global Water Resources: Vulnerability from Climate Change and Population Growth." *Science* 289(July): 284–288.
Vörösmarty C. J., P. B. McIntyre, M. O. Gessner, D. Dudgeon, A. Prusevich, P. Green, S. Glidden, S. E. Sullivan, C. Reidy Liermann, and P. M. Davis. 2010. "Global Threats to Water Security." *Nature* 467: 555–561.
Water Education Foundation. 2015. "The 2014 Sustainable Groundwater Management Act: A Handbook to Understanding and Implementing the Law." Sacramento, CA: Water Education Foundation. http://groundwater.ucdavis.edu/files/208021.pdf. Accessed September 27 2017.
White, Gilbert F. 1945. *Human Adjustment to Floods: A Geographical Approach to Flood Problems in the United States.* Department of Geography Research Paper 29. Chicago, IL: University of Chicago.
Whitfield, P. H., and Pomeroy, J. W. 2016. "Changes to Flood Peaks of a Mountain River: Implications for Analysis of the 2013 Flood in the Upper Bow River, Canada." *Hydrological Processes* 30(25): 4657–4673.
Wisner, Ben, Piers Blackie, Terry Cannon, and Ian Davis. 2004. *At Risk: Natural Hazards, People's Vulnerability and Disasters* (Second Edition). London, UK and New York: Routledge, p. 51.
World Bank. 2012. "Thai Flood 2011: Overview." http://documents.worldbank.org/curated/en/677841468335414861/pdf/698220WP0v10P106011020120Box370022B.pdf. Accessed September 22 2017.
World Water Development Report. 2016. "Water and Jobs." http://unesdoc.unesco.org/images/0024/002439/243938e.pdf. Accessed September 22 2017.
Zimmerman, Kim Ann. 2015. "Hurricane Katrina: Facts, Damage & Aftermath." Live Science. August 27 2015. https://www.livescience.com/22522-hurricane-katrina-facts.html. Accessed September 27 2017.

3

Why Is Uncertainty a Game Changer for Water Policy and Practice?

3.1 Introduction

Climate change and its uncertain impacts on water systems has raised awareness of the limitations of current planning and management practices in the water sector. Global climate modeling and regional impact assessment are plagued with severe uncertainties stemming from natural variability, uncertain assumptions about future emissions, differences in how global models are structured, and what climate processes they include and exclude. The same model with slightly different initial conditions can give quite different results. Therefore, it is difficult to use them to predict future climate at local and regional scales. This chapter articulates the problem of climatic uncertainty, identifies other relevant uncertainties associated with demand and regulation, and discusses the challenges they present for water planning and decision-making.

P. Gober, *Building Resilience for Uncertain Water Futures*,
https://doi.org/10.1007/978-3-319-71234-5_3

3.2 Risk and Uncertainty

Some people use the terms risk and uncertainty interchangeably, but they have different implications for environmental assessment and decision-making. The concept of risk[1] includes both the probability of occurrence and the consequences associated with an event. One important component of risk is a probability distribution that sets out the likelihood of a particular event occurring, whereas uncertainty characterizes situations in which these probability distributions are unknown (Wynne 1992). Standard methods of risk analysis come from a variety of disciplines including game theory, operations research, hydrologic engineering, and statistical decision theory. Decision scientists have applied them to policy problems for many decades. These methods are relevant to problems in which system behavior is predictable, and practitioners accept probability functions as the basis from which to assess risk and make rational decisions.

Uncertainty refers to problems where we may know certain features of the system but do not know the probability distributions that describe the range of possible outcomes (Wynne 1992). Uncertainty implies use of the precautionary principle. Investments (e.g., infrastructure, insurance, land use planning, emergency management training) are committed now to safeguard against potentially catastrophic future outcomes for problems where probabilities are unknown or are unknowable. The precautionary principle has been the basis for adaptive natural resource management, using an abundance of caution when future outcomes are uncertain.

Problems of deep uncertainty are a subset of uncertainty problems in which there are profound disagreements about the key drivers that shape future outcomes, the probability density functions that quantify outcomes, and how to value the gains and losses from particular results (Lempert et al. 2003). Typically, decision-makers, citizens, and scientists co-produce scenarios or future storylines, and the consequences of these narratives are the basis of scientific modeling, public discussion, and decision-making.

3.3 Climatic Uncertainty

3.3.1 Climate Change and Water

Climate change will alter global patterns of hydrology, including the occurrence and magnitude of floods and droughts, the availability of water resources, patterns of demand, and physical processes that transfer water from the land to the atmosphere. The most recently published Intergovernmental Panel on Climate Change (IPCC) Assessment Report (AR5) concluded that increasing carbon dioxide in the atmosphere results in warmer temperatures, sea level rise, and local and regional changes in precipitation (Jiménez Cisneros et al. 2014). Water is the main mechanism for delivering climate-based impacts to society. Potential water-related impacts include more extreme events (e.g., droughts, floods, sea level rise, and cyclones in coastal regions), water quality problems, and changing availability of regional and local water supply and demand. Arid and semi-arid regions (in the neighborhood of 30° north and south of the equator) are especially prone to climate change impacts due to the nature of atmospheric circulation processes. Warmer water, more intense rainfall, and longer periods of low flow in mid-latitude regions are expected to reduce water quality and human and environmental health, change the reliability of water services, and present new challenges for the management of sea and flood defenses and water infrastructure and operations.

The highest rates of warming will occur in higher latitudes, and they are already having effects on cold-region climates and water systems. In Northern Canada, the glaciers have receded, permafrost has melted, and the active layers over permafrost have increased, with profound consequences for regional water supply systems, infrastructure, and the density of vegetation communities. The region is highly vulnerable to climate change impacts with additional pressure from natural resource extraction, hydropower generation, irrigation demand, and population growth (DeBeer et al. 2016).

Agriculture accounts for almost 70% of water withdrawals worldwide, and thus the impacts of climate change are of particular concern to farmers and economic activities related to food production (United Nations Food and Agricultural Organization 2014). Water volumes and the seasonality of stream flows also affect hydropower production and the energy sector more generally. Rising temperatures alter seasonal flows, increase flood and drought frequencies, and raise reservoir evaporation. Effects on municipal services vary by region and locale. They include increased evaporation from lakes and reservoirs, decreased water storage capacity, more frequent and intense droughts that increase the need for artificial storage, greater likelihood of algal blooms, increased pollutant concentrations in areas with dry conditions, increased storm water run-off, and sea level rise (Jiménez Cisneros et al. 2014).

3.3.2 Sources of Uncertainty

There are uncertainties associated with the IPCC's model results and the regional and local water assessments that derive from them. The IPCC has attempted to translate what is an uncertainty problem into risk-based estimates of potential impacts by quantifying the range of climate model results and giving weight to expert agreement about the impacts of changing climate conditions. Predictions of future climate rely on a suite of models developed since 1990 to represent atmospheric dynamics. They contain different statistical parameterizations to represent unresolved physical processes such as the formation of clouds and precipitation, ocean mixing due to wave processes, sea-ice interactions, and land surface interactions. The IPCC science community has rigorously represented uncertainties in the model results, although these differing results have bedeviled public debate and limited easy integration of model results into decision-making frameworks. Despite 25 years of assiduous effort to reduce uncertainty in climate model predictions, uncertainty is a reality for scientific practice and the societal institutions that rely on a science-based perspective to inform decision-making.

The IPCC differentiates three types of uncertainties associated with future assessments (Cubasch et al. 2013, pp. 138–142) (Fig. 3.1). (1) Internal or natural variability includes the natural fluctuations in climate that occur on an inter-annual basis (Hawkins and Sutton 2009). Climate

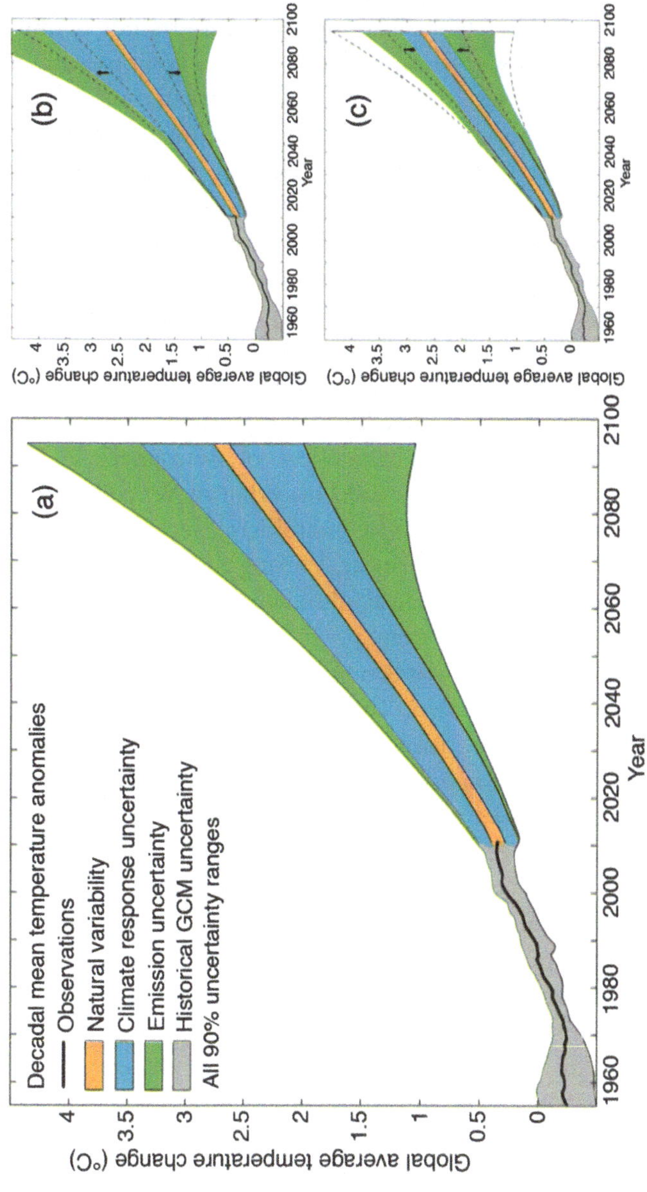

Fig. 3.1 Types of uncertainties in IPCC models. IPCC

varies naturally on nearly all time and spatial scales, and quantifying precisely natural variability is challenging. This uncertainty is unlikely to change over time. (2) Model uncertainty (climate response uncertainty) refers to the uncertainties introduced by how particular models represent the physical and biogeochemical aspects of the climate system as well as the model's response to external forces, such as greenhouse gas (GHG) emissions. In other words, each model represents the climate system differently, and their results reflect different conceptualizations of atmospheric physics. Model or climate response uncertainty represents the variation introduced by the fact that the models differ in terms of their mathematical make-up and emphasis on certain atmospheric processes. As models become increasingly sophisticated over time and include more complicated processes and feedbacks, such as the releasing of GHGs from melting permafrost or the fertilizing effects of atmospheric CO_2 on vegetation, this is likely to increase model uncertainty. When, in an editorial in *Nature Climate Change*, IPCC climatologist Kevin Trenberth (2010) warned of the problem of performing "cutting edge science in public," he was referring to uncertainty that would increase as new research tools are developed and models incorporate previously unknown processes.

(3) Scenario or emissions uncertainty stems from the fact that future GHG emissions are unknown. They relate to future lifestyles, levels of development, and policies relevant to GHG emissions. These types of uncertainties are likely to grow over time. A recent example is the uncertainty surrounding the Paris Climate Agreement negotiated by 195 countries in 2015. This agreement calls for individual countries to contribute a particular share of emission reductions to ensure that global average temperature does not rise more than 2°C above preindustrial levels. It is not yet clear how and whether countries will meet their future emissions targets of the voluntary, non-binding agreement. Moreover, US President Donald J. Trump claims that the Paris Climate Accord puts "onerous energy restrictions" on the US, and he has pulled the country out of it (White House 2017). What this means for future global emission levels is uncertain at this time.

The very best atmospheric scientists in the world have been unable to resolve the uncertainties associated with model and future emissions levels. Moreover, the political processes surrounding emissions reduction are in flux. Science is unlikely to reduce model uncertainties any time in the near future, and politicians are unlikely to reduce emissions uncertainties fast enough to influence current and mid-term local and regional water decisions. Building infrastructure, altering the built environment, implementing conservation programs, and changing human behavior can take decades.

Climate change is not a traditional risk problem where probabilities are well known and the system well defined. It is a deep uncertainty problem that requires different types of decision analysis, new forms of scientific discovery, greater emphasis on the precautionary principle, more input from the public about what is socially acceptable, and governance structures that favor experimentation, flexibility, and adaption over control.

3.3.3 Coping Ranges and Non-stationarity

Modern societies have adapted to climate extremes through engineering works, building codes, floodplain maps, warning systems, and financial instruments such as insurance programs and contingency funds. These investments address the natural variability in climate and hydrological systems. Knowledge about natural variability has allowed societies to develop coping ranges and use these for risk-based analysis (Fig. 3.2). The core of the human coping range contains acceptable outcomes for human health, property damage, and social disruption. Approaching the critical thresholds, outcomes become more negative, but most people can accommodate them. Beyond the critical thresholds, damages and losses are no longer tolerable and the potential for catastrophic loss increases (Willows and Connell 2003).

Extreme conditions in a changed climate will fall outside of society's ability to cope. Two events of similar magnitude may have different outcomes, depending on their timing (Fig. 3.3). Adaptive action after the first event reduces the impacts of the second event in the top diagram (Karl et al. 2008). In the bottom diagram, a second extreme event occurs before the community has completely recovered from and adapted to the previous one. It has a total impact in excess of what would have occurred in

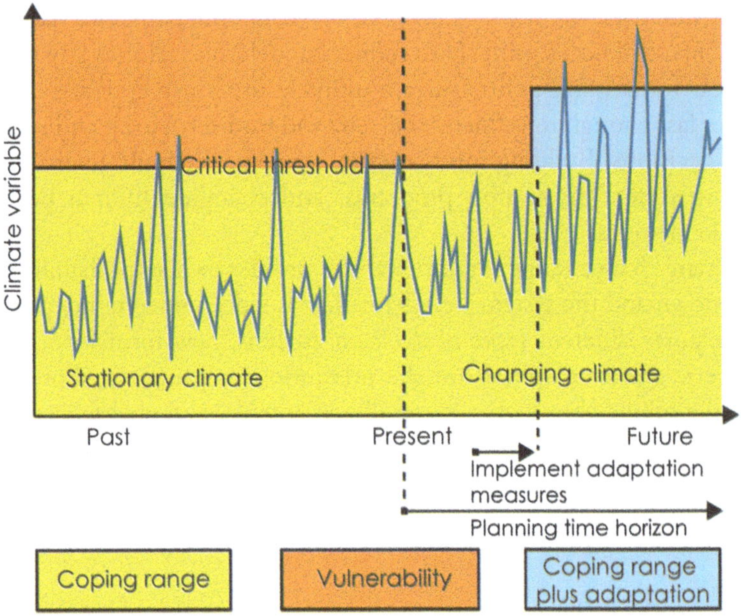

Fig. 3.2 Extreme conditions in a changed climate will fall outside of society's ability to cope. Centre for Ecology & Hydrology, UK

isolation. That is what happened when a series of hurricanes made landfall in Florida in 2004. They occurred so close together in time and space that communities did not have enough time to recover before the next event occurred (Pielke et al. 2008). A similar response occurred during the 2003 heat wave in Europe. Anomalous hot and dry conditions affected Southern and Central Europe between June and mid-August 2003, raising temperatures by 3–5°C. The warm conditions in June lasted throughout the month, but did not cause excess mortality until the second episode occurred between August 1 and 13 when temperatures were more than 7°C above normal (Vandentorren and Empereur-Bissonnet 2005). In the latter case, people had not yet processed the first event and figured out how to protect elderly people living alone from heat stress.

Hydrological uncertainties affect water resource management. Traditional management training and practice is based on the assumption of stationarity—the idea that natural systems operate within a known envelope of variability. It implies that probability density distributions

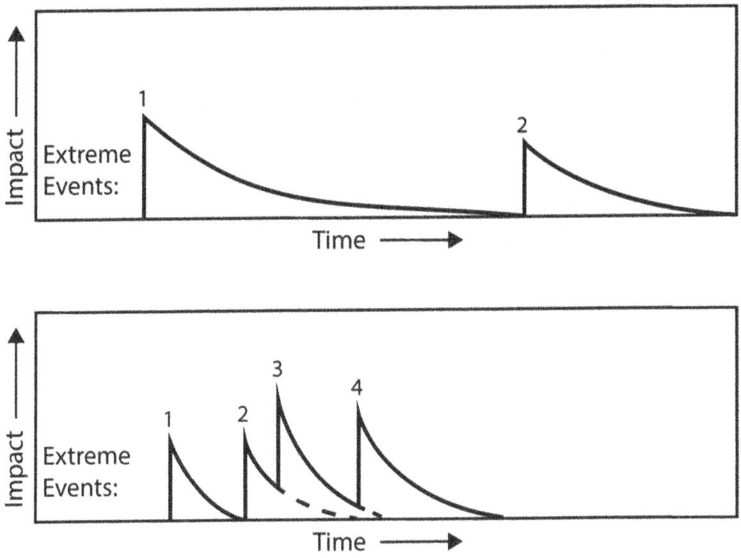

Fig. 3.3 Extreme events result in greater impact as they increase in both frequency and intensity. US Climate Change Research Program

adhere to empirical conditions (e.g., peak discharge and flood peak) in the historical record. These density distributions guide infrastructure construction and operations, floodplain mapping, and water resource planning. In a very influential 2008 article in *Science*, Milly et al. declared that "Stationarity is dead: Whither water management" (Milly et al. 2008, p. 573). The end of stationarity stems from human-induced climate change that has altered precipitation means and extremes, evapotranspiration, and rates of river discharge. The authors argue that the science community needs to find new ways to represent uncertainty in non-stationary water systems. In other words, we have entered a new normal when experiences of the past are no longer guides to plans for the future.

Non-stationarity has profound consequences for the management community where accepted training and practice are organized around probability density distributions, risk analyses, the search for optimal solutions, and rational decision-making (e.g., cost/benefit analyses, optimization modeling). Probabilistic solutions are not well suited to non-stationary conditions, and water utilities should look for new methods of water planning (Quay 2010). Using probabilistic frameworks to represent non-stationary systems exposes the water sector to vulnerabilities that are relevant to these uncertainties (Hall and Borgomeo 2013). Stirling (2010) urges the consideration of a wider range of approaches and different conceptualizations of uncertainty to look for alternatives that are robust in the face of an unpredictable future. A robust decision is, by definition, one that is insensitive to variable future conditions (Lempert and Schlesinger 2000). Conventional infrastructure designs tend not to be robust to climate change. They are, in fact, planned using probability distributions for stationary climatic and hydrological conditions.

Wilby and Dessai (2010) criticized the use of global climate model-based assessments to inform local and regional water decisions based on concerns about uncertainty. The traditional top-down method of climate adaptation first downscales climate projections from global climate models under a range of emission scenarios, then feeds results into hydrological models of stream flows, and finally uses these products in water resource models to predict future water supply. A "cascade of uncertainties" expands at each step to introduce a wider range of future conditions. An alternative, bottom-up strategy for assessment starts at the local level addressing known vulnerabilities. UK water-supply vulnerabilities can be reduced by converting abstraction licenses to a time limited basis so they are more flexible to uncertainties of climate change, increasing the connectivity of water supply systems to encourage sharing, increasing tariffs to improve efficiency, asking abstractors to reduce withdrawals during drought conditions, and providing incentives for high-efficiency fixtures and appliances. Globally, vulnerability reduction includes flood-proofing homes and businesses, moving vulnerable populations from hazardous areas, monitoring for safe drinking water, reducing groundwater depletion, and improving governance (Wilby and Dessai 2010).

3.4 Uncertainty and Transition Management

Rigid and specific goal setting, difficulty working across sectors and agencies, inflexible decision rules, fixed allocation systems, short time horizons, and fragmented governance inhibit capacity to manage uncertainty.

Yet, these are common characteristics of water management throughout the world. Uncertainty challenges not only the resource itself but also the assumptions that have guided administrative and operational practices in the water sector. Ecology faces a similar challenge in biodiversity conservation, although the latter approach is shifting away from persistence of existing conditions to managing inevitable ecological change to prevent biodiversity loss and degradation in ecosystem services (Stein et al. 2013).

Camacho (2009) articulated three elements of a more transformative approach to resource management at the federal level in the US. First, make decisions in natural resource management more proactive than reactive. Proactive decisions occur before environmental damage occurs. They avoid damage, but risk making investments that turn out later to have been unnecessary. Reactive decisions minimize investments in measures that ultimately prove unnecessary, but expose society to high-level damage and potential system collapse. Second, emphasize decisions that produce co-benefits. Investment in green infrastructure, for example, mitigates climate-related urban heat island effects and enhances flood control but also promotes healthier, more pedestrian-oriented lifestyles. Controlling CO_2 emissions from less household energy use produces less air pollution but also reduces lifestyle illnesses such as heart disease and obesity (Harlan and Ruddell 2011). Third, change the focus of scientific and public discussion from climate change per se to the societal structures needed to manage climate uncertainty. In Europe, Huitema and Meijerink (2010) raised the issue of transition management in the water sector—how to restructure administrative systems for radical change without waiting for shock events to occur. These systems include policy entrepreneurs who develop new ideas for policy and governance, political coalitions that sell new ideas and exploit windows of opportunity, and administrative and public venues for coalition building and decision-making.

Social scientists have explored the organizational structure of water utilities in North America and found them highly focused on reliability—water for all sectors and all purposes at all times (Lach et al. 2006). Water engineers, economists, and lawyers dominate these agencies, supporting infrastructure planning and operations, pricing, and allocation decisions. They build infrastructure and develop organizational practices to smooth out the natural fluctuations in water to make water systems more predictable and thus more reliable. Organizational culture prioritizes local knowledge and so-called craft skills at the expense of innovation and experimentation and operates largely outside the public realm. The response to growing water stress in North America is to spread the risks associated with rapidly changing physical and human environments across a wider range of stakeholders and organizations while protecting core organizational norms in the water sector.

3.5 Demand Uncertainties

Water demand is also fraught with uncertainties (Jiménez Cisneros et al. 2014). Because agriculture accounts for almost 70% of all water withdrawn from the global water system, changes in food consumption and production patterns affect global water demand. Demand is also influenced by population growth, economic development, and changing lifestyles. The growing demand for energy increases water demand because water and energy production are highly interrelated. Energy production uses water for a variety of purposes, including the extraction of raw materials. In Canada, for example, bitumen (oil sands) recovery requires, on average, 2.8 barrels of freshwater for every barrel of bitumen produced (Canadian Oil & Natural Gas Producers 2016, p. 43). Nuclear power and other thermal plants use water for cooling; biofuels require water to cultivate crops; and hydropower generation uses water to power turbines.

Many urban water utilities in North America have incorporated climate change into forecasting models. In a study of future demand for six geographically diverse North American water utilities (Colorado Springs, Colorado; Durham, Ontario; Boston, Massachusetts; Las Vegas, Nevada; San Diego, California; and Tampa Bay, Florida), Keifer et al. (2013) found that municipal water demand was highly sensitive to regional differences in climate and to the variability in actual weather conditions. Estimated demand increases for the six cities under different climate change scenarios ranged from 1% to 23% by 2055 and from 2% to 45%

by 2090, depending on the climate scenario, region, and specific utility. Water-sensitive demands existed in all the cities they studied and accounted for a substantial portion of total demand in most regions and a majority of total demand in some areas and seasons. Some cities and scenarios produced increases in demand that were larger than projected future declines in water use, suggesting that gains in net water resources from conservation may not account for net losses from climate change impacts. Particularly significant is the increasing role of outdoor use in influencing demand portfolios (Gober et al. 2016; Quay 2015).

Until recently, planning for future water demand in North American cities assumed that per capita demand was constant and that future demand would mirror past demand. In other words, the stationarity assumption applied to both demand and supply. There was relatively little incentive to understand the determinants of demand and the effects of particular policy decisions on it. The 1992 US Energy Policy Act, implemented in 1994, mandated high-efficiency fixtures and appliances in new and retrofitted homes. To comply, major plumbing manufacturers nationwide began producing low-volume toilets, urinals, showerheads, and faucets. Over the past several decades, these products have slowly become more common as new homes are constructed and older homes retrofitted with new appliances. Coomes et al. (2010) found that the introduction of low-flow toilets, showers, and washing machines had a significant impact on residential water use, accounting for a decline of about 16% in average daily use between 1990 and 2010. On a per capita basis, water demand in Seattle declined 36% between 1990 and 2009 (Flory 2013). Per capita residential water use in Phoenix fell from 143 gallons (541 liters) to 106 gallons (401 liters (City of Phoenix Water Services Department 2014)).

Assuming that most of the decline in per capita water use has come from more efficient indoor fixtures and appliances, future efficiencies will need to come from outdoor water use. There is a high level of uncertainty associated with outdoor water conservation because it is not a technical problem like requiring better indoor water fixtures. Outdoor demand is difficult to predict because it is climate-sensitive, highly peaked in hot summer months (Fig. 3.4), and linked to cultural attitudes and personal

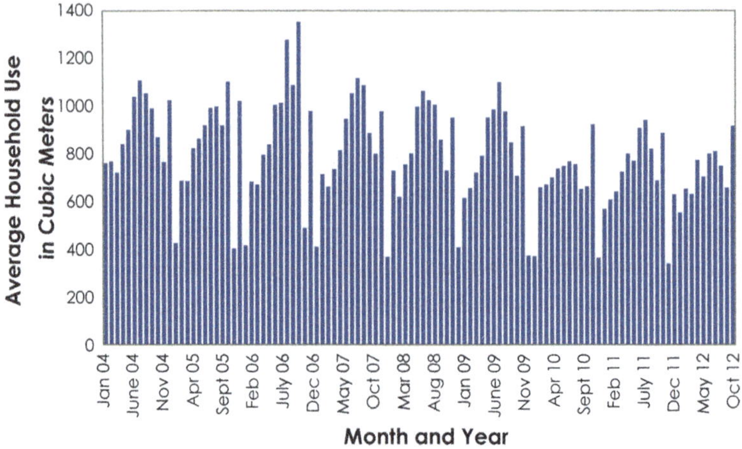

Fig. 3.4 Outdoor water in Saskatoon is highly peaked in the summer season. City of Saskatoon Water Services Department

behaviors. Also uncertain is how households will react to policy signals to change behavior. In a survey of water use in 700 single-family homes across 10 California water agencies, DeOreo et al. (2011) showed that 87% of homes irrigate their yards, 54% of homes that irrigate are doing so in excess, and 62% of excess use is occurring on just 15% of lots. Effective conservation programs thus must target the relatively small number of over-irrigators, many of whom use automatic timers and lawn service contracts to maintain their gardens.

It is unclear how households will respond to calls for use reductions, and social scientists are only beginning to understand the behavioral aspects of water conservation. Kenney et al. (2004), for example, tracked households in six Colorado Front Range cities during drought conditions in the summer of 2002. They found that voluntary restrictions had little impact, mandatory restrictions reduced water use, and the greatest savings occurred in cities with the most aggressive and stringent mandatory restrictions. Maggioni (2015) in Southern California found that outdoor water restrictions curbed per capita water use, but water rates and subsidies to install water-saving devices did not. Wickman et al. (2016) found that low-income households in single-family detached dwellings in North Carolina were more sensitive to price than high-income households. The level of response across income categories to prescriptive, non-price policies such as voluntary and mandatory restrictions on outdoor watering was uniform. They concluded that the burden of price increases would fall more heavily on poor households, while mandatory irrigation restrictions would have a more equitable impact, inducing more outdoor water conservation among high-income, high-water-use irrigators, found to be most responsible for wasteful practices and insensitive to price.

Reliable predictions of per capita water demand require better monitoring of household water use, more accurate population estimates to predict the denominator in per capita use, and more knowledge about the psychology, economics, and sociology of water demand (Quay 2015; Hogue and Pincetl 2015). There is growing awareness in the urban water sector that declines in per capita water use of the past 20 years are unlikely to continue into the next 20 years as the big technology gains from high-efficiency fixtures and appliances are unlikely to translate to outdoor water where behavior and lifestyles are key in determining use. Future urban water demand is as uncertain as future water supply (Quay 2015).

3.6 Policy and Regulatory Uncertainties in Phoenix

Policy and regulatory uncertainties can also complicate water planning. Weak enforcement of Arizona's 1980 Groundwater Management Act (GMA) leaves some Phoenix-area municipalities and residents vulnerable to actions of neighboring communities via their shared groundwater resource.

Metropolitan Phoenix, with a 2016 population of 4.5 million residents, enjoys an enviable location at the base of two large watersheds of the Salt and Verde Rivers in Central Arizona. Surface water from these systems was the basis for early historical agricultural development and later large-scale urbanization (Fig. 3.5). The Salt and Verde Rivers deliver approximately one million acre-feet (1.23 billion m^3) annually to the urban region. Construction of the Central Arizona Project (CAP), a 336-mile aqueduct that delivers water from the Colorado River to the Central

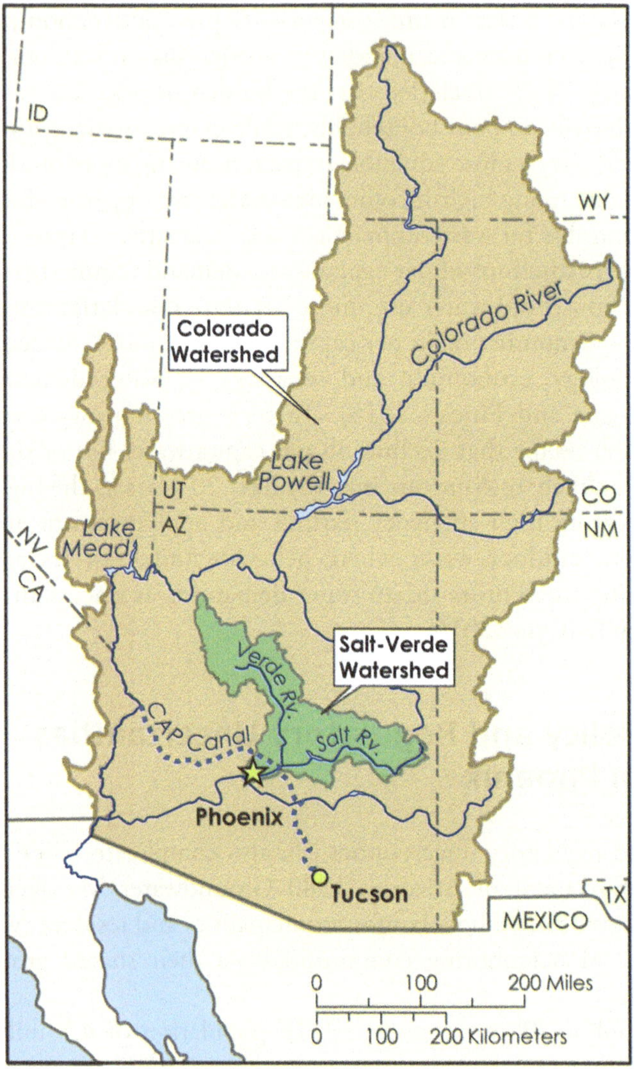

Fig. 3.5 Metropolitan Phoenix has a large hydraulic reach

Arizona cities of Phoenix and Tucson, in the 1980s augmented surface water supplies. CAP delivers approximately 1.5 million acre-feet (1.85 billion^3) of the State of Arizona's Colorado River water allocation to Phoenix and Tucson. In addition to these two surface water sources, the Phoenix area uses groundwater to support development. Together the three water sources (Salt and Verde, CAP, and groundwater) have been diverse enough and plentiful enough to support large-scale agricultural production and post-war urban expansion in this desert region (Gammage 1999; Gober 2006).

The ability to draw both surface water and groundwater was critical to Phoenix's early development; it allowed farmers to extract from aquifers during drought periods. Rapid post-World War II growth and unsustainable groundwater practices during the 1950s, 1960s, and 1970s led to significant overdrafts, the need to pump at deeper elevations, land subsidence, aquifer compaction (and reduction in storage space), decline in water quality, and structural damage to roads and canals. In 1980, the State of Arizona used approximately five million acre-feet (6.167 billion m^3) of groundwater annually, twice the estimated sustainable level (Hirt et al. 2008).

In 1980, the Arizona State Legislature passed legislation to address unsustainable groundwater use. The GMA called for gradual reductions in agricultural irrigation, increases in conservation by urban water users, and the close monitoring of groundwater extraction in so-called Active Management Areas (AMAs). Within these AMAs, new housing estates were required to demonstrate a 100-year assured water supply from renewable sources.

Fragmented water governance and political pressure from developers weakened this common-sense visionary plan for groundwater management. Water resource decisions are in the hands of 119 local public and private providers, with 32 of them accounting for more than 90% of current regional demand (Fig. 3.6). Each provider has water rights that vary by source, based on seniority of historical claims and use. They develop their own conservation policies, tariff structures, and long-term plans. They have unequally diverse and secure supply portfolios. Pressure to accommodate future demand is disproportionately in communities at the urban fringe that depend almost exclusively on groundwater. Assuming a suburban-style pattern of future growth, the central city and

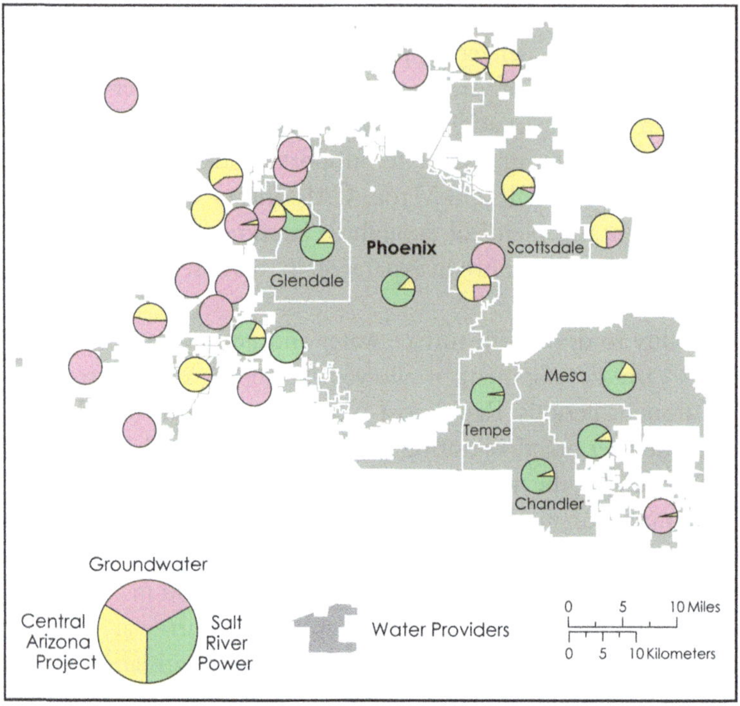

Fig. 3.6 Phoenix area water providers

its inner suburbs can meet current and future demand with their historical water rights and groundwater credits, while some outlying suburban communities are stressed to accommodate future growth and meet the state's 100-year assured supply rule from groundwater sources.

Political and legal pressure from local private developers and communities enabled development to continue by ignoring groundwater hydrology. Critics of Arizona's GMA argue that the 100-year assured supply rule amounts to a "taking" of their land rights in the sense that landowners would be unable to profit from the sale or development of their lands. Claims of "takings" also come from new suburbs well positioned geographically to capitalize on their moment of growth. The state relented on the 100-year assured supply rule and formed the Central Arizona Groundwater Replenishment District (CAGRD) in 1993. The CAGRD allows developers to use excess Colorado River surface water to meet their requirement for a 100-year assured supply. Today, the District purchases Colorado River water from the Central Arizona Project and uses it to replenish the aquifer, thus meeting the requirement for a 100-year assured supply. In recent years, the state counted between 300,000 and 400,000 acre-feet (370 million m^3 and 493 million m^3) of water annually from the Colorado River to replenish the aquifer and prevent further overdraft (Arizona Department of Water Resources 2017).

There are two major problems with the CAGRD. First, it is not at all clear that there will be unused water from the Colorado River for the next 100 years. Upstream from Phoenix on the Colorado River, water in Lake Powell was at 50% capacity and 39% of capacity in Lake Mead in June 2017. Second, there is a geographic disconnection between where the water enters the aquifer and where it leaves to support CAGRD-approved development. Surplus surface water replenishment locations are near the Central Arizona Project Canal (the source of Colorado River Colorado River water), while urban-fringe communities pump groundwater from adjacent wells (Hirt et al. 2008). Ultimately, homeowners will pay the costs of this arrangement. They are assessed a fee for the purchase of water that may grow over time with increasing scarcity. Water rates are already diverging across local water providers. Tempe, an urban-core community with senior surface water rights, has substantially lower water consumption rates than Buckeye, a rapidly growing urban-fringe community wholly dependent on groundwater (Fig. 3.7). Also significant is the potential for unsustainable, unregulated groundwater practices in urban-fringe communities to undermine the long-term supplies of neighboring communities that have planned and managed groundwater more judiciously.

3.7 Conclusions

Climate is not the only or even most important source of uncertainty facing water managers in Phoenix or elsewhere. There are also uncertainties associated with demand and the regulatory environment. The State of Arizona's failure to regulate groundwater adequately introduces uncertainties for local water managers and new home buyers. These uncertainties require more attention to the social and political aspects of decision-making

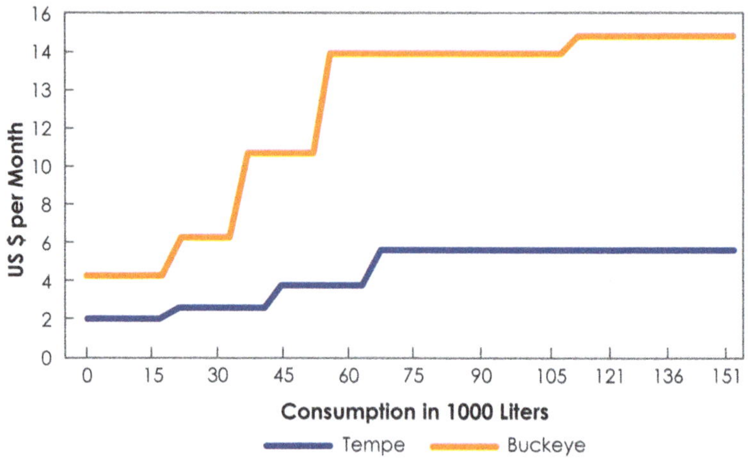

Fig. 3.7 Water consumption rates for Tempe and Buckeye

and to the relationship of water to other resource sectors. Chapter 4 argues that water's relationships with energy, food, and land present hidden vulnerabilities for economy, society, and the environment.

Notes

1. Risk has different meanings in different communities of science and practice. The flood-risk hydrological community uses risk to represent the probability of occurrence and the consequences associated with an event. Risk factors into efforts to evaluate alternative infrastructure designs and operations of existing designs. The hazards community is more likely to see the risk of harm from natural events as socially constructed and linked to social, economic, and political processes that influence how hazards affect people with different intensities.

References

Arizona Department of Water Resources. 2017. "Phoenix AMA Summary Budget." http://www.azwater.gov/AzDWR/WaterManagement/Assessments/default.htm. Accessed August 14 2017.

Camacho, Alejandro E. 2009. "Adapting Governance to Climate Change: Managing Uncertainty Through a Learning Infrastructure." *Emory Law Review* 59: 1–77.

Canada's Oil & Natural Gas Producers. 2016. "Canada's Oil Sands." http://www.google.com/url?sa=t&rct=j&q=&esrc=s&source=web&cd=4&ved=0ahUKEwip7-Pd2qfUAhUhilQKHfLzCjAQFggzMAM&url=http%3A%2F%2Fwww.capp.ca%2F-%2Fmedia%2Fcapp%2Fcustomerportal%2Fpublications%2F296225.pdf&usg=AFQjCNF_ZZGGoZtQMCVARm2c71_jJafv0Q&sig2=_ZPgEj9ZB-kivkppn_H9Yg. Accessed June 14 2017.

City of Phoenix Water Services Department. 2014. "Water Production and Water Use, 1990–2013." Obtained through public record request.

Cubasch, U., D. Wuebbles, D. Chen, M. C. Facchini, D. Frame, N. Mahowald, and J-G Winther. 2013. Introduction. In: *Climate Change 2013: The Physical Science Basis. Contribution of Working Group I to the Fifth Assessment Report of the Intergovernmental Panel on Climate Change*, edited by T. F. Stocker, D. Qin, G-K. Plattner, M. Tignor, S. K. Allen, J. Boschung, A. Nauels, Y. Xia, V. Bex and P. M. Midgley. 119–158. Cambridge, UK and New York: Cambridge University Press.

DeBeer, Chris M., Howard S. Wheater, Sean K. Carey, and Kowk P. Chun. 2016. "Recent Climatic, Cryospheric, and Hydrological Changes over the Interior of Western Canada: A Review and Synthesis." *Hydrology and Earth System Sciences*, 20:1573–1598, https://doi.org/10.5194/hess-20-1573-2016.

DeOreo, William B. with Leslie Martien, Matthew Hayden, Andrew Funk, Michael Kramer-Duffield, Renee Davis, James Henderson, Bob Raucher, Peter Gleick, and Matt Herberger. 2011. "California Single-family Water Use Efficiency Study." Boulder, CO; Aquacraft. http://www.aquacraft.com/wp-content/uploads/2016/04/CalSF-Water-Study-Report-Body-120811.pdf. Accessed September 13 2017.

Flory, Bruce. 2013. "Why We Are Here." Presentation to Urban Water Demand Roundtable," Tempe, Arizona, April 18 and 19, 2013. https://d3dqsm2futmewz.cloudfront.net/docs/dcdc/website/documents/2_BruceFlory_WhyAreWeHere-Flory130418.pdf?x58848. Accessed August 27 2017.

Gammage, Grady. 1999. *Phoenix in Perspective: Reflections on Developing the Desert.* Tempe, AZ: Herberger Center for Design Excellence, College of Architecture and Urban Design, Arizona State University.

Gober, Patricia. 2006. *Metropolitan Phoenix: Place Making and Community Building in the Desert.* Philadelphia, PA: University of Pennsylvania Press.

Gober, Patricia, Ray Quay, and Kelli L. Larson. 2016. "Outdoor Water Use as an Adaptation Problem: Insights from North American Cities." *Water Resources Management* 30(3): 899–912.

Hall, John, and Edoardo Borgomeo. 2013. "Risk-based Principles for Defining and Managing Water Security." *Philosophical Transactions of the Royal Society* 371: https://doi.org/10.1098/rsta.2012.0407.

Harlan, Sharon L., and Darren M. Ruddell. 2011. "Climate Change and Health in Cities: Impacts of Heat and Air Pollution and Potential Co-benefits from Mitigation and Adaptation." *Current Opinion in Sustainability* 3(3): 126–134.

Hawkins, Ed, and Rowan Sutton. 2009. "The potential to Narrow Uncertainty in Regional Climate Predictions." *Bulletin of the American Meteorological Society* 90: 1095–1107.

Hirt, Paul, Annie Gustafson, and Kelli L. Larson. 2008. "The Mirage in the Valley of the Sun." *Environmental History* 13(July): 482–514.

Hogue, Terry S. and Stephanie Pincetl. 2015. Are You Watering Your Lawn? *Science* 348(6241): 1319–1320.

Huitema, Dave, and Sander Meijerink. 2010. "Realizing Water Transitions: The Role of Policy Entrepreneurs in Water Policy Change." *Ecology and Society* 15(2): 26.

Jiménez Cisneros, B. E., T. Oki, N. W. Arnell, N. W. G. Benito, J. G. Cogley, P. Döll, T. Jiang, and S. W. Mwakalila. 2014. *Freshwater Resources.* In: *Climate Change 2014: Impacts, Adaptation, and Vulnerability. Part A: Global and Sectoral Aspects.* Contribution of Working Group II to the Fifth Assessment Report of the Intergovernmental Panel on Climate Change, edited by C. B. Field, V. R. Barros, D. J. Dokken, K. J. Mach, M. D. Mastrandrea, T. E. Bilir, M. Chatterjee, K. L. Ebi, Y. O. Estrada, R. C. Genova, B. Girma, E. S. Kissel, A. N. Levy, S. MacCracken, P. R. Mastrandrea, and L. L. White, 229–269, Cambridge, UK and New York: Cambridge University Press.

Karl, Thomas R., Gerald A. Meehl, Christopher D. Miller, Susan J. Hassol, Anne M. Waple, and William L. Murray. 2008. *Weather and Climate Extremes in a Changing Climate.* Synthesis and Assessment Product 3.1 Report of the U.S. Climate Change Science Program and the Subcommittee on Global Change Research. https://www.climatecommunication.org/wp-content/uploads/2012/01/climateextremes.pdf. Accessed September 22 2017.

Kenney, Douglas S., Roberta A. Klein, and Martyn P. Clark. 2004. "Use and Effectiveness of Municipal Water Restrictions During Drought in Colorado." *Journal of the American Water Resources Association* 40(1): 77–87.

Kiefer, Jack C., John M. Clayton, Benedykt Dziegieleski, and James Henderson. 2013. "Analysis of Changes in Water Use Under Regional Climate Change Scenarios." Denver, CO, Water Research Foundation Accessed on June 20, 2017 at: http://www.waterrf.org/Pages/Projects.aspx?PID=4263.

Lach, Denise H., Helen Ingram, and Steve Rayner. 2006. "Maintaining the Status Quo: How Institutional Norms and Practices Create Conservative Water Organizations." *Texas Law Review* 83(7): 2027–2053.

Lempert, Robert J., and Michael E. Schlesinger. 2000. "Robust Strategies for Abating Climate Change." *Climatic Change* 45(3–4): 387–401.

Lempert, Robert J., Steven W. Popper, and Steven C. Bankes. 2003. *Shaping the Next One Hundred Years: New Methods of Quantitative, Long-term Policy Analysis*. Santa Monica, CA: Rand.

Maggioni, Elena. 2015. "Water Demand Management in Times of Drought: What Matters for Water Conservation?" *Water Resources Research* 51: 125–139.

Milly, P. C. D., Julio Betancourt, Malin Falkenmark, Robert M. Hirsch, Zbigniew W. Kundzewicz, Dennis P. Lettenmaier, and Robert J. Stouffer. 2008. "Stationarity Is Dead: Whither Water Management?" *Science* 319(5863): 573–574.

Paul A. Coomes, Barry D. Kornstein, Thomas D. Rockaway, Joshua A. Rivard. 2010. "North America Residential Water Usage Trends." *Proceedings of the Water Environment Federation* 2010 (9): 6488–6500.

Pielke Jr. Roger A., Joel Gratz, Christopher W. Landsea, Douglas Collins, Mark A. Saunders, and R. Rade Musulin. 2008. Normalized Hurricane Damages in the United States: 1900–2005. *Natural Hazards Review* 9(1): 29–42.

Quay, Ray. 2010. "Anticipatory Governance: A Tool for Climate Change Adaptation." *Journal of the American Planning Association* 76(4): 496–511.

Quay, Ray. 2015. "Planning for Demand Uncertainty in Integrated Water Resource Management." *Journal of the American Water Works Association* 107(2): 32–41.

Stein, Bruce A., Amanda Staudt, Molly S. Cross, Natalie S. Dubois, Carolyn Enquist, Roger Griffis, Lara J. Hansen et al. 2013. Preparing for and Managing Change: Climate Adaptation for Biodiversity and Ecosystems. *Frontiers in Ecology and the Environment* 11(9) (2013): 502–510.

Stirling, Andy. 2010. "Keep It Complex." *Nature* 438:1029–1031.

Trenberth, Kevin. 2010. "More Knowledge Less Certainty." *Nature Reports Climate Change* 4(February): 20–21.

United Nations Food and Agriculture Organization. 2014. "Water Withdrawal." http://www.fao.org/nr/water/aquastat/infographics/Withdrawal_eng.pdf. Accessed June 22 2017.

Vandentorren, S., and P. Empereur-Bissonnet. 2005. "Health Impact of the 2003 Heat-Wave in France" In *Extreme Weather Events and Public Health Responses*, edited by W. Kirch, B. Menne, and R. Bertolinni, 81–88. Berlin: Springer.

White House Press Office. 2017. "Statement by President Trump on the Paris Climate Accord." https://www.whitehouse.gov/the-press-office/2017/06/01/statement-president-trump-paris-climate-accord. Accessed September 27 2017.

Wickman, Casey J., Laura O. Taylor, and Roger H. von Haefen. 2016. "Conservation Policies: Who Responds to Price and Who Responds to Prescription?" *Journal of Environmental Economics and Management* 79: 114–134.

Wilby, Robert L. and Surage Dessai. 2010. "Robust Adaptation to Climate Change." *Weather* 65(7): 180–183.

Willows, Robert and Richenda K. Connell (Eds.). 2003. *Climate Adaptation: Risk, Uncertainty and Decision-making*. UKCIP Technical Report. UKCIP, Oxford. http://www.ukcip.org.uk/wp-content/PDFs/UKCIP-Risk-framework.pdf. Accessed August 14 2017.

Wynne, Brian. 1992. "Uncertainty and Environment Learning: Reconceiving Science and Policy in the Preventive Paradigm." *Global Environmental Change* 2(June): 111–127.

4

Hidden Vulnerabilities in the Water-Energy-Land-Food (WELF) Nexus

4.1 Introduction

Traditional assessments focused on water resources alone ignore vulnerabilities related to water's connections to food, energy, and land. The international trade in food includes embedded water, sometimes with unintended consequences. Water problems may appear in unexpected ways because of these interdependencies. In 2016, drought conditions in Thailand and flooding in Pakistan and China led to a drop in production of 63 million mega tons of grain, leading to export bans on rice and grain and elevated grain prices on the world market (Lewis 2017). In 2012, Hurricane Sandy downed power lines, flooded underground substations, and interrupted power supplies in the Northeastern US, closed eight nuclear power facilities, and interrupted electrical service to some eight million customers in 21 states (US Department of Energy 2013). Changing land use preferences that favor irrigated lawns over traditional Mediterranean gardens led to water crises in Barcelona and several other Mediterranean cities (Domene and Saurí 2006). In complex systems where water connects to energy, food, and land, vulnerabilities can occur

P. Gober, *Building Resilience for Uncertain Water Futures*,
https://doi.org/10.1007/978-3-319-71234-5_4

in surprising and exaggerated ways. This chapter examines interrelationships among water, energy, land, and food (WELF) sectors.

4.2 Virtual Water

Geographer Tony Allen first coined the term "virtual water" in 2003 to convey that agricultural products require substantial quantities of water, and the import of food can ameliorate water shortages (Allan 2003). Allan used the concept of virtual water to argue that water-scarce countries in the Middle East are better off importing food than using their meager water resources to produce it. Theoretically, the process of international trade negotiates the supply and demand of both crops and water resources, but some food exports are the result of local or regional water use that is not factored into the market, as we saw in the Aral Sea problem earlier (Chap. 2).

Critics of virtual water argued that it is confusing to assert that water exchange occurs in the trade of agricultural products. Moreover, it is difficult to measure for public policy purposes. Nonetheless, virtual water, as a concept, entered public discourse in the early 1990s, and the Stockholm Water Prize in 2008 honored Tony Allan for drawing attention to the fact that water is rooted in the production and trade of food, industrial products, and consumer goods. The production of a cup of coffee requires 140 liters of water to grow, produce, package, and ship the beans, equivalent to the amount of water used by the average person daily in England for drinking and household needs (Water Footprint Network 2017).

At an operational level, it has proved difficult to value precisely the water portion of exported agricultural products and to differentiate between blue (surface and groundwater) and green (rainfall and soil moisture) water. Measurement remains a cause of disagreement, but there is now general acceptance that trade in agricultural products involves a de facto trade in water. For water-stressed countries, importing water-intensive products, while exporting less water-intensive commodities, relieves pressure on domestic water resources (Konar et al. 2011). From a policy perspective, Allan observed that virtual water is a potentially powerful tool to alleviate short-term water deficits—far more efficient, in his view, than building additional supply capacity in the form of large dams (Allan 2003).

Efforts to operationalize virtual water led to the concept of a "water footprint," the volume of water needed to produce the goods and services that are consumed by inhabitants of a place (Hoekstra and Chapagain 2007). First advanced in 2002 by Arjen Y. Hoekstra from UNESCO-IHE, the water footprint differentiated between domestic water used for the production of goods and services by the inhabitants of a country and water from imported goods and services. The external water footprint illustrates how residents of some countries are dependent on external water supplies via the international trade in food and how the growth in meat-eating diets exacerbates this dependence. In the United Kingdom, 72.3% of the water used to produce agricultural and industrial goods comes from external sources, compared to 55.2% in Germany, and 21.6% in Australia (Hoekstra and Chapagain 2007). Paterson et al. (2015) championed the idea of water footprints for cities, emphasizing that water flows into and out of cities via exchanges of goods and services. Urban policy, land planning, and water decision-making should account, in their view, for these virtual water flows.

Meats and meat products require more water per kilocalorie than grains, cereals, fruits, and vegetables (Table 4.1). Per ton of product, meat generally has a larger water footprint than crop products. The same is true for the water footprint per calorie of food intake. The average water footprint per calorie of beef is 20 times larger than the footprint for cereals and starchy roots. The water footprint per gram of protein for milk, eggs, and chicken meat is about 1.5 times larger than for pulses. For beef, the water footprint per gram of protein is six times larger than for pulses. All other animal products have larger water footprints per gram of fat when compared to oil crops. From a freshwater resource perspective, it is more efficient to obtain calories, protein, and fat through crops than animal products (Water Footprint Network 2017). These data show the importance of diets that favor more meat products in increasing stress on global water resources and the potential to save water by shifting to less meat-based diets.

Table 4.1 Water footprint of selected animal and food crops

	Liter/ kilogram	Liter per kilocalorie	Liter per gram of protein	Liter per gram of fat
Sugar crops	197	0.69	0.0	0.0
Vegetables	322	1.34	26	154
Starchy roots	387	0.47	31	226
Fruits	962	2.09	180	348
Cereals	1644	0.51	21	112
Oil crops	2364	0.81	16	11
Pulses	4055	1.19	19	180
Nuts	9063	2.63	139	47
Milk	1020	1.82	31	33
Eggs	3265	2.29	29	33
Chicken	4325	3.00	34	43
Butter	5553	0.72	0.0	6.4
Pork	5988	2.15	57	23
Sheep/goat meat	8763	4.25	63	54
Beef	15,415	10.19	112	153

Source: Mekonnen, M.M. and Hoekstra, A.Y. 2010. "The green, blue and grey water footprint of farm animals and animal products," Value of Water Research Report Series No.48, UNESCO-IHE. Accessed on June 9

The water footprint concept also reveals how dependent particular countries are on imported food, and thus on imported and embedded water. Hoekstra and Chapagain (2007) estimated the total water footprint for countries (1997–2001). They differentiated between those where the national consumption of food and industrial products is highly dependent on water like the US (2483 m^3 per capita of water per year), Canada (2049), Russia (1858), and Germany (1545) and those with smaller personal water footprints, such as India (980), Bangladesh (896), and China (720). The water footprint has become a talking point for discussions about meat-eating diets, the international trade in virtual water, strategies for saving water, and the security of global supply chains.

4.2.1 Virtual Water Flows in China

Investigators also use the water footprint of regions to quantify flows of embedded water in the internal trade in food. Zhuo et al. (2016), for example, showed that the national average water consumption per capita in China fell by 23% between 1978 and 2008, owing to increasing crop yields internally and increasing imports of crops from countries with relatively small water footprints. Internal agricultural patterns shifted significantly from production in the water-rich South to the water-scarce North. The total harvested area of major crops grew by 16% in the North but decreased by 13% in the South. Regional virtual water flows needed to support this redistribution of production and consumption within China were, in fact, larger than the international flows in virtual water. Food flow patterns changed from South to North before 2000 to North to South after 2000. The Northeast shifted from being a net importer to being a net exporter of rice. Continuation of these trends puts increasing pressure on the North's already scarce water resources. The ongoing South-North Water Transfer Project (SNWTP) alleviates pressure to a certain extent. The Middle Route of the SNWTP, which became operational after late 2014, transfers three billion m^3 of blue water (surface and groundwater) per year to the North to support agriculture, with the aim of increasing irrigated land by 600,000 hectares. The government's plan to expand irrigated agriculture by using the transferred water for irrigation stimulates crop exports from the North and blue water transfer from South to North. The blue water supply through the SNWTP thus does not significantly reduce pressure on water resources in the North, however. Long-term solutions to the North's water problems are more likely to involve changing dietary preferences in China away from meat and importing more food than moving more water around to support grain production in the North.

Zhuo et al. (2016) projected virtual water flows in China to 2030 and 2050, using results of climate models, projections of population growth, estimations of possible expansion in harvested crop area, potential crop yields through technological development, and hypothetical changes in diet. Results are subject to large uncertainties about future climate, implementation of policy tools, and behavioral responses to these policies. Nonetheless, they do offer insight into the range of outcomes and the future of the country's water resources. Climate change model-based scenarios increase the water footprint of crops produced in the dry regions of Northwest China. In most areas of the Northeast, the green (soil moisture from precipitation) and blue water footprints per ton of crops increase by more than 40% as ET increases and precipitation declines. Diet is a major uncertainty and one with significant impacts on the future water footprint. Reducing meat and meat-based components to half of current levels would reduce China's future water footprint for food consumption by 44% in 2050. The Chinese government's goal of achieving self-sufficiency in food production, while reducing its water footprint, depends on continued technological development, effective cropland management, low population growth, and a behavioral change in favor of less meat in the diet.

There are two key messages here for global water resource management. The first is that water security is not only about water but also about food trade and dietary patterns. Just as changing diets and consumption patterns have stressed water systems, they can also reduce stress. A second takeaway message is that water is not fully valued in the international trade in food. As a result, importing countries such as the UK can externalize the true costs of high water footprints. Similarly, China can disguise regional water shortages and continue unsustainable internal transfers because the inter-regional trade in virtual water is invisible to the market. This invisibility makes it difficult to monitor, predict, and avoid disruptions to supply chains that result from water shortages. It is more difficult to anticipate surprises in systems like these because virtual water flows are not generally integrated into planning schemes and risk assessments.

4.2.2 NAFTA Trade in Avocados

Avocados are now the second largest export crop for Mexico, after tomatoes, thanks in part to the steady rise in price and the growing popularity of the Mexican fruit among American consumers. Demand has doubled over the past 15 years. Mexico grows some 80% of the avocados consumed in the US. Mexico has ideal growing conditions for avocados, with the central state of Michoacán producing 85% of the exports (Fariza 2017).

Avocados are water intensive. Thirty-two gallons of water is required to produce one pound of Mexican avocados. The fact that avocado production occurs in regions that are prone to drought means that farmers must tend their orchards to ensure not only the economic viability of avocado production but also the availability of sufficient water for human consumption and irrigation of other crops. Reduction in tariff barriers resulting from the North American Free Trade Agreement (NAFTA) encouraged rapid increases in Mexican avocado production with little regard for virtual water and unsustainable water use (Kruskal 2016). Alejandro Macías, specialist in regional economics at Mexico's University of Guadalajara, warns that avocado production in some areas "is leading to water shortages; we are exporting virtual water to other countries" (Fariza 2017). Current media discussions in the US about the pros and cons of NAFTA rarely include the large-scale export of water from Mexico to the US via the avocado trade.

4.3 Energy-Water Nexus

The linkage between energy and water use means that a shortage or disruption in one resource can trigger problems for the other. The production of energy requires large volumes of water for cooling and turning turbines. Total withdrawals for power generation accounted for 45% of total US water withdrawals (Maupin et al. 2010). In Europe, 44% of abstracted water is used for cooling, most of it being returned to rivers (Walsh et al. 2015). Water abstraction, treatment, desalination, and distribution are highly dependent upon low-cost energy. In Europe, it takes 0.37 kWh/m^3 to withdraw water from a river, 0.48 kWh/m^3 to pump groundwater, 0.62–0.87 kWh/m^3 for wastewater treatment, 1.0–2.5 kWh/m^3 for wastewater reuse, and 8.5 kWh/m^3 for desalination (Walsh et al. 2015). Thus, the high cost of water purification favors the segregation of use and treatment of water for intended use. The transportation and treatment of water, disposal of wastewater, and the energy used to heat and consume water accounted for nearly 20% of the total electricity and 30% of non-power plant-related natural gas consumed in California (California Energy Commission 2005).

Climate change has ramifications for the energy sector via its heavy reliance on water. Limited water available for cooling at thermoelectric facilities affects power plant utilization. Increased evaporation rates or changes in the timing of snowpack affect the timing of water available for hydropower. Decreased water availability affects biofuel production (US Department of Energy 2015). Recent drought conditions in the American West lowered water levels in Lake Mead behind Hoover Dam, straddling the border between Arizona and Nevada (Figs. 4.1 and 4.2). The Hoover Dam power plant was the largest facility of its kind when it began operations in 1936, commissioned to produce 2.080 megawatts of electrical power via 17 generators (US Bureau of Reclamation 2017).

The current capacity for the Hoover Dam power plant has fallen to 1.592 megawatts as levels in Lake Mead declined due to drought conditions in the West. As lake levels decrease, there is less pressure as the water falls down the intake toward the turbines below. The power plant mainly now provides energy only for peak demand during daytime hours (Kuckro 2014). Major customers of Hoover Dam's energy are the Metropolitan Water District of Southern California, the State of Nevada, the State of Arizona, the City of Los Angeles, Southern California Edison, and ten other downstream cities. A recently completed redesign enables the dam to produce energy more efficiently under low-flow conditions, but falling levels in Mead jeopardize the long-term production of inexpensive (and

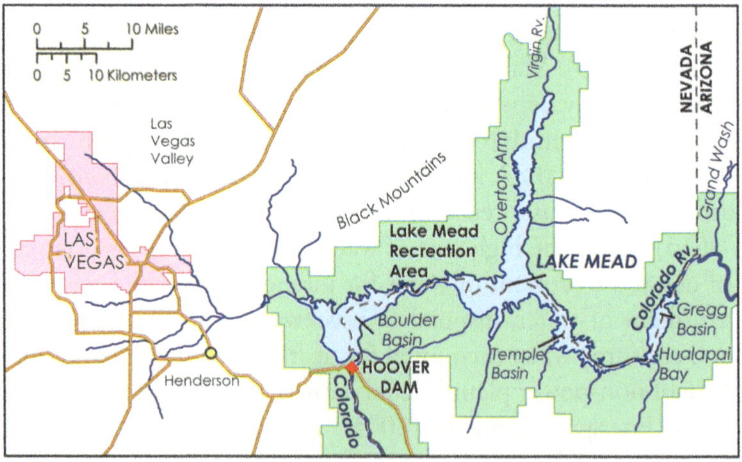

Fig. 4.1 Colorado River at Lake Mead adjacent to Las Vegas

Fig. 4.2 Hoover Dam and hydroelectric facilities. US Department of the Interior/ Bureau of Reclamation

zero carbon) hydropower. Low lake levels also threaten water supplies to the City of Las Vegas, just 25 miles from Hoover Dam, which receives 90% of its water supplies from the reservoir.

Concern about the energy-water nexus is by no means limited to North America and Europe. Rogers (2009) assessed future energy and water needs in the Greater Mekong Sub-region in Vietnam, Laos, Cambodia, Thailand, and Myanmar and found considerable challenges in meeting future water demands, driven largely by increasing urbanization and changing lifestyles. Existing fully allocated water supply systems face increasing water demands related to increasing food and energy demands from a larger urbanized population. Some countries, such as the Lao People's Democratic Republic, diverted water and implemented water-saving technologies in both the water and energy sectors, but these efforts are unlikely to be enough to support the growing food, energy, and water needs of their populations. Expensive methods for meeting water demand, such as recycling and desalination, also increase the embedded energy demanded. Rogers estimates that urban water demand will triple by 2030, and electricity's demand for water could represent as much as 71% of the total urban water supplies in some areas. While capacity for electrical power generation is growing, water supply is not increasing fast enough to keep pace with increasing demands from the energy sector.

Energy-water interconnections exacerbate the problems of intermittent water and energy supplies in many cities of the developing world. Malik (2002) called attention to the effects of erratic water supplies on energy use in Delhi, India. To accommodate for restricted, unreliable, and insufficient water supplies delivered at inadequate pressure, many urban households reorganize daily routines around the availability of water supplies, store water in buckets and tanks, and outfit their residences with tube wells that draw underground water for domestic uses. Such strategies lower the water table and expose households to water quality problems if the water is not treated. Households also install voltage correction systems, such as mini generators, and above-ground tanks for storage and treatment. Often, they buy power from electricity retailers at a higher price than they would pay the local utility. In the end, households use excess energy to cope with the problem of intermittent water supplies. It would take less energy to improve the reliability of the water supply than what is used to correct the problems associated with the intermittent supplies. Better coordination of the two sectors would benefit both.

Although harmonization of the two sectors is an obvious solution to nexus-related problems, there are significant differences in the way energy and water systems are structured and managed in most parts of the world. Energy is often a national resource managed for market conditions, while water is a local resource, run according to a larger set of community values.

Energy production in the US occurs in power plants and is distributed to customers via the North American Electrical Grid, a complex system of transmission lines, substations, and transformers. The stability of the grid requires coordination of different components. Some utilities generate all the electricity they sell, using just the power plants that they own. Others purchase electricity from other utilities, power marketers, and independent producers. The retail structure varies regionally and includes not-for-profit municipal electrical utilities, cooperatives, private for-profit utilities, power marketers, and a few federally owned power authorities (e.g., Tennessee Valley Authority, Bonneville Power Administration). Prices vary by the type of fuel, the operational costs of power plants, the costs of maintaining the transmission and distribution systems, and regulations that vary from one state to another (Energy Information Administration 2016). The price of energy reflects general market conditions of supply and demand, although regulation is stronger in some places than in others (Overturf 2012).

The US energy sector aggressively responded to the need to deal with water- and climate-related emergencies. Assessment occurs at the federal level, and focus is on decreasing water use at power plants, increasing non-renewable energy use that is not dependent on water availability, and increasing efficiencies in production and demand. Efforts are also under way in some cities to curtail peak demand. Planting trees and green roofs lowered peak use by 2–3% in New York City. Development of more water- and energy-efficient fixtures and appliances, mandated by the 1992 US Energy Policy Act, also reduced both energy and water use. The Department of Energy estimates that adoption of energy and water conservation standards resulted in annual water savings of 1.5 trillion gallons in 2010, and projects a cumulative water savings of more than 51 trillion gallons by 2040 (Meyers et al. 2011). Proponents from the energy side stress the need for increasing efficiencies in both sectors, with the energy sector gaining from water efficiencies and vice versa. They have gone so far as to suggest that water should be managed more like energy, stressing markets, technology, research and development, national-level initiatives, and metrics and data analysis (US Department of Energy, 2014; Walsh et al. 2015).

The idea of water as a commodity for which users compete in a national-level market does not gibe with the realities of water governance in the US, however. Water allocations result from complex and multi-scale legal frameworks that implement rules and regulations governing water use. The price of water only rarely reflects market conditions, although markets in some water-scarce areas (e.g., California and Australia) allow buyers and sellers to trade legal water allocations. Pricing serves a range of objectives, such as equity, public health, environmental efficiency, financial stability, simplicity, public acceptability, and transparency (Arbués et al. 2003). In many urban areas, the tariff structure is a way to recover the costs of water provision, fund capital improvements, signal scarcity, and distribute the cost of operations across different types of users. Often, water decisions reflect the community-based nature of the resource and the community's social values and risk management preferences.

4.3.1 Bhopal Residents Rely on Tankers for Water Supply

Rapid urbanization resulted in unprecedented rise in water demand in Bhopal, India. Overuse of groundwater led to a drop in water levels of up to 50% during the past decade. During summer months, many areas receive water at low pressures once every day for up to 30 minutes and in some cases every other day. A modern city dubbed "City of the Lakes" now has residents in new housing estates that are almost completely dependent on small tanker deliveries of water with no end in sight. A significant amount of energy is used in transporting and pumping (into and out of tankers of) water. It also takes many hours of human labor used for hauling, storing, rationing, and contingency planning to compensate for intermittent and unreliable 30-minute daily supply of water (*Times of India* 2017).

The hardest hit neighborhoods in Bhopal are dependent on the Upper Lake, which is a major water source for the city. Lake levels fell by 1.35 feet (0.3 m) in less than one month in summer 2016. Many communities do not have pipelines and depend upon boreholes for their water supply. As the water levels dropped, they received intermittent supplies from the city and, when those were unavailable, purchased delivered water from trucks. Residents complain that they are in a constant struggle to get drinking water. Water comes even in the middle of the night causing residents to choose between sleep and drinking water. These problems occur in neighborhoods adjacent to the Bhopal School of Planning and Architecture and Design, the Indian Institute of Science Education and Research, and the National Institute of Fashion Technology, distinguished institutions in a modern city of more than three million residents (*Times of India* 2017; *Hindustan Times* 2016).

Differences in modes and scales of governance stymie cross-sector coordination between energy and water resources. The operational linkages between the two sectors are huge, translating weaknesses (and strengths) from one sector to the other. The examples discussed above are but the tip of the iceberg of vulnerabilities that stem from water's heavy reliance on energy and vice versa (e.g., Klein et al. 2005; Stillwell et al. 2010).

4.4 Land-Water Nexus

4.4.1 Urban Water and Land

Decisions about urban land use have de facto consequences for water use and increasing water scarcity. In most cities, land use decisions are independent of water service provision (Bates 2012). Water supply has typically been the purview of water managers driven by the imperative to support future land and industrial development and population growth. Land planning traditionally regulates development to avoid conflicts; community values strongly shape land use regulation. Planners, as a matter of course, participate in community discussions and integrate public views into their decision-making. Water planning has been subordinate to land use planning in the sense that it is the job of the water manager to obtain water to meet the demands of population growth but rarely to question where or how much growth occurs. Failure to coordinate across sectors leads to perceptions of natural water shortages when, in fact, uncoordinated land use decisions pre-ordain these shortages. Hill and Polsky (2007) coined the term "suburban drought" to characterize water shortage conditions created by preferences for irrigated lawns, backyard pools and spas, and water features such as fountains and misters, typical of sprawling suburban development.

In *Lawn People*, Robbins (2007) noted that modern lawns are the fastest growing landscape in the US, requiring more time and care than any other managed landscape. They are part of the North American mentality and economy. Lawns function in a socioecological system consisting of the people who care for them, the businesses that profit from them, and the institutions that regulate them. Cultural preferences for an idealized golf course-like appearance lead to heavy application of chemicals and widespread use of artificial watering systems, even in humid regions. Water for outdoor use, including lawn irrigation, now accounts for at least one-third of residential water use in the US. This figure is even higher in drier regions of the South and West (US Environmental Protection Agency 2013). Cultural preferences for lawns and land development patterns to support them lead to seasonal peaks in water use with commensurate needs that match peak demand (Gober et al. 2016).

Del Moral Ituarte and Giansante (2000) and Kallis and Coccossis (2003) linked water crisis and drought alerts in southern European metropolitan areas to the growth of low-density suburbs. Domene and Saurí (2006) studied the effects of urban sprawl on water consumption in Barcelona, where the traditionally compact Mediterranean city has given way to a diffuse urbanized region dominated by low-density housing at the urban fringe. Analysis of household surveys matched with local water records revealed that detached houses at the urban fringe with turf grass landscape treatments and gardens larger than 200 m^2 used water at 10 times the per-household rate of homes in high-density, urban-core residential settings. Moreover, these low-density detached homes have highly peaked consumption patterns that stress water supply infrastructure during hot, dry summer months. Traditional water conservation programs focused on pricing, taxation, and cutbacks during drought periods miss the deeper structural causes of water insecurity in Barcelona—the growing preference for irrigated lawns and gardens that are embedded in patterns of urban sprawl. House-Peters et al. (2010) found that affluent neighborhoods with larger homes and more outdoor space in the Portland suburb of Hillsboro use more water per household, and thus are more drought sensitive, than older, denser neighborhoods in the urban core.

These studies show why it will be difficult to reduce outdoor water use. Many residents are highly invested in their watered landscapes and the features that surround homes. There are legacy effects built into most urban landscapes such that people make de facto water decisions (pools, spas, lawns, fruit trees, etc.) when they purchase their homes. Also significant is the rise in homeowner associations (HOAs), corporations formed by real estate developers for the purpose of marketing, managing, and selling homes and lots as residential subdivisions (Lerman et al. 2012). Reducing outdoor water use is highly dependent on their CC&Rs (covenants, codes, and restrictions) that dictate irrigation regimes, types of plants, and turf grass coverage; land use planning; and personal tastes.

4.4.2 Urban Heat Island (UHI) Effects

UHI effects are the most obvious impact of urbanization on local-scale weather and climate (Oke 1982). UHI effects create an urban area that is warmer than the surrounding rural countryside (Unger 2004). UHI effects are most prominent at night when excess heat stored in urban surfaces during the daytime is released into the atmosphere. UHI intensity is determined by building density and height-width ratio, road and traffic density, building and surface materials, vegetation type and density, sky-view factor (exposure of urban surfaces to the sky), and local and regional synoptic weather conditions (Brazel et al. 2007; Hart and Sailor 2009).

Phoenix has significant UHI effects due to post-World War II large-scale urban land development. Summer nighttime temperatures are as much as 5°C warmer in the urban core than in the fringe, this in a desert city that was not considered fit for human habitation until the popularization of air conditioning after World War II (Brazel et al. 2007). The pattern of irrigated agriculture and urban water use affects the intensity of UHI effects in Phoenix. Agricultural lands on the urban fringe on the city's western edge and in-town neighborhoods rich in irrigated crops, trees, and shrubs are significantly cooler than heavily urbanized neighborhoods with many impervious surfaces. A matched set of remotely sensed images show land use and vegetation from a Landsat satellite image at 30 m resolution in April 2017 (Fig. 4.3a) and nighttime temperatures from an earlier Aster image at 90 m resolution shot in March 2007 (Fig. 4.3b). It is much hotter in the commercial core and adjacent neighborhoods than in agricultural lands on the southwestern urban fringe and in isolated inner-core neighborhood that still rely on flood irrigation to water lawns and trees.

One way to manage UHI effects in Phoenix is to plant and maintain irrigated trees and shrubs that give off moisture, shade the surface, and promote nighttime cooling. Irrigation water supports this cooling strategy. Gober et al. (2010) studied the effects of vegetated landscaping on outdoor water use and nighttime cooling in 10 neighborhoods in Phoenix, where UHI effects are very pronounced and have been linked to human health problems and deteriorating comfort (Harlan et al. 2006; Ruddell et al. 2010). The 10 neighborhoods included industrial zones dominated by impervious surfaces and buildings, residential zones with native desert landscape treatments, and residential zones with lush vegetated cover. Results showed that the presence of vegetated surfaces accelerates nighttime cooling, but there are limits beyond which adding more watered landscapes produces little additional cooling. This point occurs when the proportion of trees and grasses exceeds about 45% of the land surface.

The relationship between vegetated surfaces and nighttime cooling begs the question of how much water it would take to cool a desert city like Phoenix or a semi-arid city like Portland on summertime evenings. At the heart of this question is the sustainability trade-off between water conservation and temperature amelioration. A study of this trade-off showed that vegetative fraction (percentage of land cover in grasses, trees, and shrubs) slightly moderates daytime heating rates but significantly enhances nighttime cooling (Fig. 4.4a–d). Water application is most efficient in arid neighborhoods where adding vegetation produces significant cooling. Adding vegetation to already highly vegetated neighborhoods has only marginal impacts on cooling capacities (Gober et al. 2012).

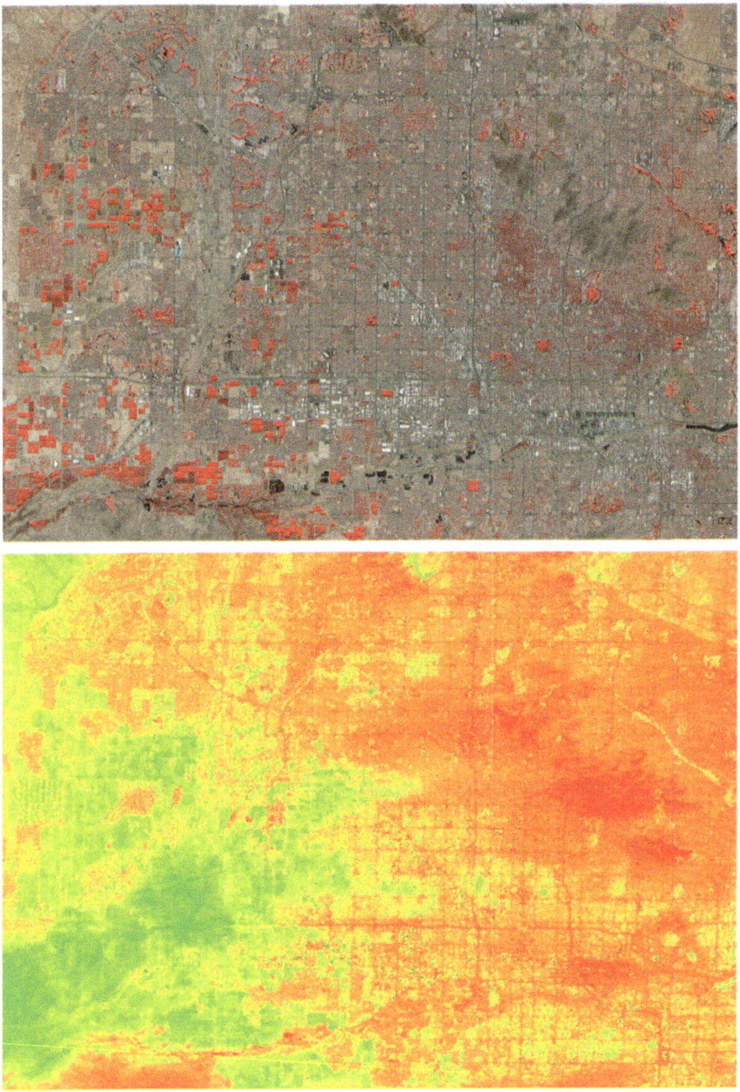

Fig. 4.3 (**a**) Phoenix land cover. LANDSAT false color satellite image, 30 m resolution, April 26, 2017, and (**b**) nighttime temperatures, ASTER image, 90 m resolution, March 5, 2007. Soe Myint/School of Geographical Sciences and Urban Planning/Arizona State University

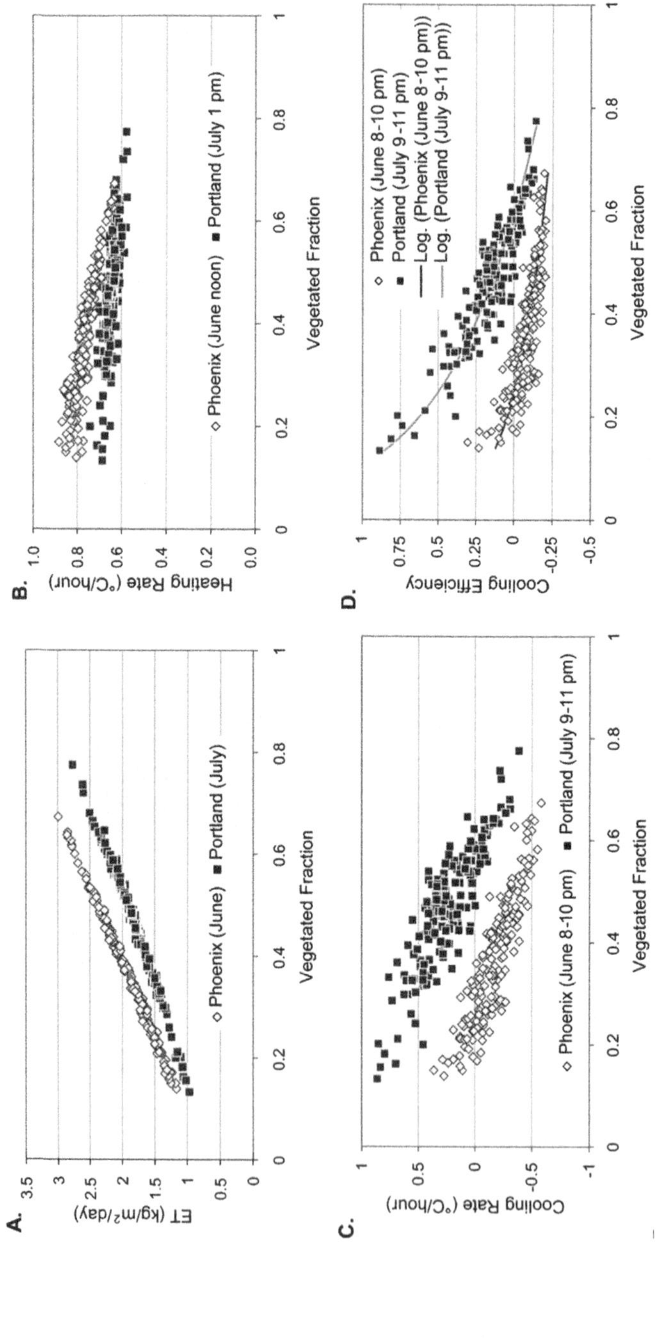

Fig. 4.4 (**a**) evapotranspiration (ET) rates, (**b**) cooling rates, (**c**) heating rates, and (**d**) efficiency rates in Phoenix and Portland

Future changes in land use in both cities affect water use and temperature management. Future scenarios include decreasing the density of vegetation (a xeriscaping scenario), increasing vegetation density (a greening scenario), and replacing impervious surfaces with built surfaces (a densification scenario). Implementing these future scenarios in Phoenix and Portland alters climate-water-land relationships in both cities. The greening scenario accelerates nighttime cooling but increases outdoor water use by about 20% in both cities. The xeriscaping scenario decreases outdoor water use by 15% in Phoenix and 13% in Portland, but reduces the capacity for heat management. Both cities warmed faster during the day and cooled more slowly at night. Most of Phoenix did not begin to cool until after 10 p.m., significant for a city that aspires to create outdoor public spaces to enliven the urban experience. The cooling rate in Portland changed from 0.24°C in the base case to 0.36°C (in other words, it continued to heat up) in the xeriscaping scenario. Both cities lost cooling efficiency under this scenario. A 15% savings in water came with a 122% loss in nighttime cooling capacity in Phoenix; in Portland a 13% savings in water resulted in a 53% loss in cooling capacity. The densification scenario produced the most efficient combination of temperature and water management conditions, producing similar improvements in nighttime cooling (to the greening scenario) but with minimal increases in water use. Replacing impervious surfaces with buildings produced gains in cooling without significantly adding to outdoor water use.

Interconnected land use and water decisions influence the livability of cities and their ability to cope with climate change. A single-minded interest in water conservation ignores water's links to land use and urban climate. Deeper insights into these linkages shed light on the difficult choices that lie ahead for land planners and water managers and the vulnerabilities that result from coordination failures.

4.4.3 Ethanol

Ethanol is an alcohol fuel distilled from plant materials such as sugar and corn. Mixed with gasoline, ethanol is a fuel for cars. In an effort to increase energy independence, the US Congress passed the Energy Policy

Act of 2005, incentivizing production of ethanol. Rapid growth of ethanol production from 6.5 billion US gallons (24,600,000 m^3) in 2007 to 13.3 billion US gallons (50,300,000 m^3) in 2012 and 14.8 billion US gallons (56,000,000 m^3) in 2015 (Renewable Fuels Association 2016) affected land use, energy production, and water use in the US. It is the quintessential nexus issue with far-reaching and long-lasting consequences for natural resource development. In 2012, Congress extended the target production of ethanol to 36 billion gallons (140,000,000 m^3) by 2022 with the Energy Independence and Security Act of 2007 (U.S. Congress 2007). The story of ethanol is one of unintended consequences in an increasingly complex resource system.

After 2005, the production of corn for ethanol grew steadily due to high gasoline prices and subsidies for the early development of domestic energy production. By the end of 2010, 88% of the gasoline used to fuel vehicles in the US blended ethanol with gasoline (Renewable Fuels Association 2011, p. 6). Ethanol production reached more than one million barrels per day in early 2017 (Fig. 4.5). Corn is the main feedstock used to produce ethanol in the US (Brazil depends on sugarcane for its ethanol production). The market significance of additional demand for corn (or sugarcane)

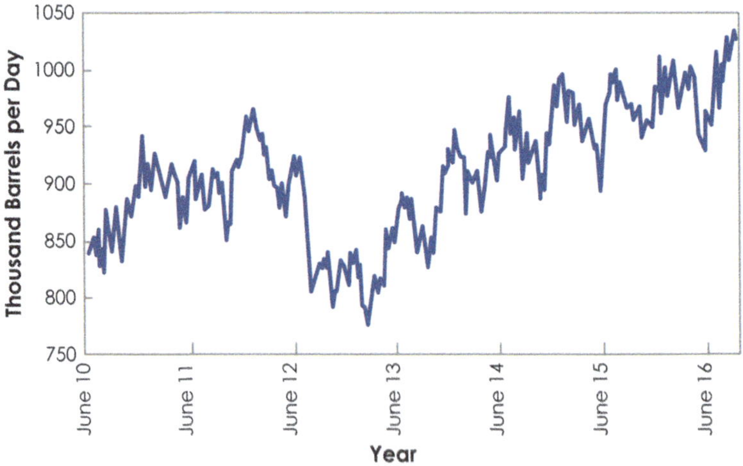

Fig. 4.5 Ethanol production in the US, 2010 to 2016. US Energy Information Administration. https://www.eia.gov/todayinenergy/detail.php?id=32152

has important implications for domestic crop and livestock production, as well as for global trade and international markets (USDA 2017).

Policy questions center on the amount of land and water used to grow the corn to produce ethanol. Also relevant is the amount of energy consumed in the production process whereby corn is converted to ethanol versus the energy in ethanol, the amount of carbon produced by ethanol production and transportation, and the effects of ethanol production on food prices. Public discussion of ethanol has centered on energy, but connections to the larger WELF nexus are also relevant, especially in light of the fact that ethanol is subsidized by the US federal government (Foley 2013; Hamelinck 2013). Critics note that the 40% of corn production in the US that is now used for ethanol rather than food raises the price of food nationally and internationally.

Corn production consumes blue water withdrawn from rivers and aquifers and green water from evapotranspiration. It contributes to the dead zone of nutrient overloading at the mouth of the Mississippi River in the Gulf of Mexico (Fig. 4.6). Donner and Kucharik (2008) used scenarios of land use and physically based models of aquatic nitrogen to evaluate the effect of the expansion of corn-based ethanol production on nitrogen levels in the Gulf of Mexico. They found that the amount

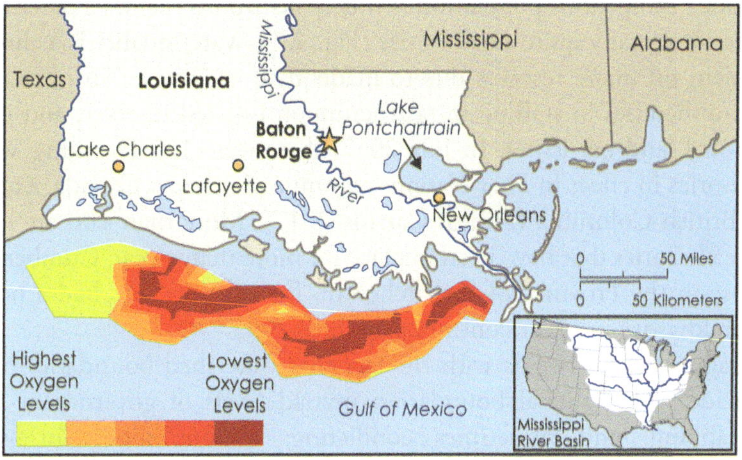

Fig. 4.6 Dead Zone in Mississippi, July and August 2015. NOAA/http://www.noaa.gov/media-release/gulf-of-mexico-dead-zone-is-largest-ever-measured

of corn cultivation required to meet the target of 36 billion gallons of renewable fuels by 2022 would increase dissolved nitrogen export by the Mississippi and Atchafalaya Rivers from 10 to 34%. Declining water quality in the Gulf of Mexico is, in other words, a by-product of policy efforts to make the US more energy independent with biofuel production.

The decision problem about ethanol is complicated when large-scale ethanol production is part of a complex WELF resource system. Public attitude studies reveal skepticism about ethanol based on fairness (high food prices in poor countries), economic impact (high food prices at home), and concerns about environmental benefits (energy costs of producing and distributing ethanol) (Delshad et al. 2010). Ethanol advocates may have a harder time selling the benefits of ethanol production and energy independence if ethanol is part of a cluster of WELF issues that implicate it in a global debate about food, land, energy, water, and the environment.

4.4.4 Source Water Protection on Canadian First Nations Lands

Canada's Indigenous population is widely distributed across a patchwork of reserves, many quite small in size (Fig. 4.7). Water quality is a chronic problem on many reserves due to inadequate investment in infrastructure, difficulties in staffing infrastructure on isolated reserves, and jurisdictional land use issues. In July 2017, there were 150 Drinking Water Advisories in effect in First Nations communities across Canada, excluding British Columbia (Health Canada 2017). More than 100 are long-term advisories that have been in place for more than a year. Cumberland House in the Province of Saskatchewan, for example, has had a boiled water advisory in effect continuously since 2003.

Fragmentation begins with the fact that watershed boundaries rarely coincide with political boundaries; myriad levels of government have overlapping, and sometimes conflicting, responsibilities for water management. The 1930 Constitution Act between Canada and each of its western provinces assigns responsibility for land and resources, including

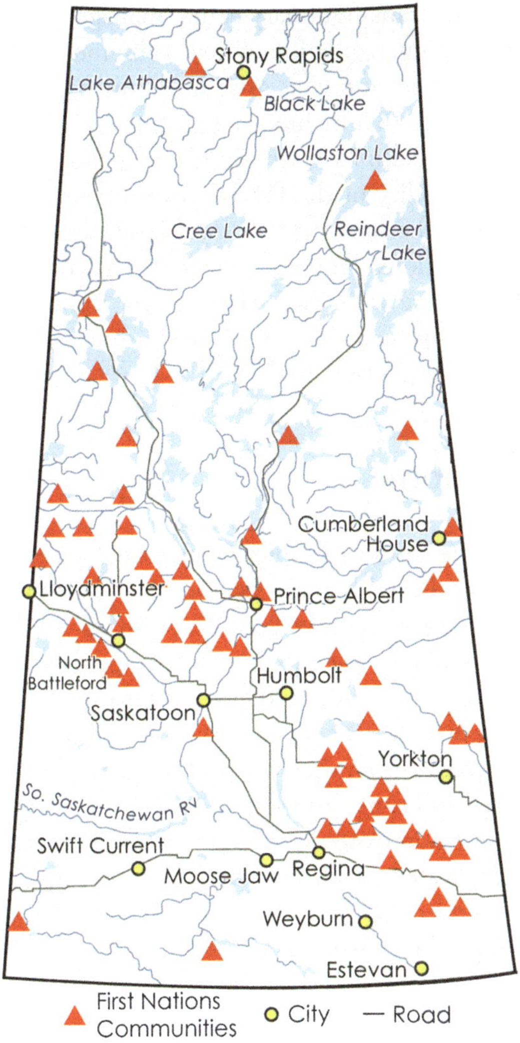

Fig. 4.7 First Nations communities in Saskatchewan, Canada

water, to the provinces (Partners for the Saskatchewan River Basin 2009). As such, the provinces are responsible for infrastructure, flood forecasting, drinking water standards, water quality management, source water protection, water licenses, and fisheries regulation. The federal government has responsibility for trans-boundary issues, management of fish habitat, navigation and shipping, regulation of toxic substances, and First Nations lands. Obvious conflicts and gaps arise in the management of a multi-provincial water system that includes a large number of small, geographically dispersed First Nations reserves. Water quality on First Nations lands is the responsibility of the federal government, but protection of source waters lies at the provincial or local level through groundwater–surface water interactions, land use management, community development, and collaborative watershed partnerships. Boiled water advisories reflect the lack of coordination between authorities responsible for land and water governance (Patrick 2011).

4.5 Conclusions

This chapter ran the gamut of topics from virtual water and ethanol to the urban heat islands and boiled water advisories on Indigenous lands. The scope of these topics represents water's vast reach beyond the water sector and the potential for seemingly unrelated processes related to food trade, energy production, and land planning to have far-reaching and unexpected consequences for water. Psychologists use the term "syndrome" to characterize the intersection of a set of causal processes and outcomes in space and time. The idea of a syndrome provides a point of entry to talk about water problems as one piece of an intricate suite of interacting issues. Water thus becomes a way to discuss hidden vulnerabilities associated with biofuels, urban design, food trade, Indigenous governance, and power interruptions.

References

Allen, J. A. (Tony). 2003. "Virtual Water – The Water, Food, and Trade Nexus: Useful Concept or Misleading Metaphor? IWRA," *Water International* 28(1): 4–11.

Arbués, Fernando, Maria Ángeles Garcia-Valiñas, and Roberto Martinez-Espiñeira. 2003. "Estimation of Residential Water Demand: A State-of-the-art Review." *The Journal of Socio-Economics* 32(1): 81–102.

Bates, Sarah. 2012. "Bridging the Governance Gap: Emerging Strategies to Integrate Water and Land Use Planning." *Natural Resources Journal* 53: 61–97.

Brazel, Anthony, Patricia Gober, Seung-Jae Lee, Susanne Grossman-Clarke, Joseph Zehnder, Brent Headquest, and Erin Comparri. 2007. "Determinants of Changes in the Regional Urban Heat Island in Metropolitan Phoenix (Arizona, USA) Between 1990 and 2004." *Climate Research* 33(2): 171–182.

California Energy Commission. 2005. "California's Water-Energy Relationship." CEC-700-2005-011-SF. Sacramento: California Energy Commission. http://www.energy.ca.gov/2005publications/CEC-700-2005-011/CEC-700-2005-011-SF.PDF. Accessed September 27 2017.

del Moral Ituarte Leandro, and Consuelo Giansante. 2000. "Constraints to Drought Contingency Planning in Spain: The Hydraulic Paradigm and the Case of Seville." *Journal of Contingencies and Crisis Management* 8(2): 93–102.

Delshad, Ashlie B., Leigh Raymond, Vanessa Sawicki, Duane T. Wegener. 2010. "Public Attitudes Toward Political and Technological Options for Biofuels." *Energy Policy* 38(7): 3414–3425.

Domene, Elena. and David Saurí. 2006. "Urbanisation and Water Consumption: Influencing Factors in the Metropolitan Region of Barcelona." *Urban Studies* 43(9): 1605–1623.

Donner, Simon D. and Christopher J. Kucharik. 2008. "Corn-based Ethanol Production Compromises Goal of Reducing Nitrogen Export by the Mississippi River." *Proceedings of the National Academy of Sciences* 105(11): 4513–4518.

Energy Information Administration. 2016. "Energy Explained: How Energy Is Delivered to Customers." http://www.eia.gov/energyexplained/index.cfm?page=electricity_delivery. Accessed January 24 2017.

Fariza, Ignacio. 2017. "Is Mexico's Avocado Boom Sustainable?" *El Paiz* March 27 2017. http://elpais.com/elpais/2017/03/27/inenglish/1490612533_740839.html. Accessed September 16 2017.

Foley, Jonathan. 2013. "It's Time to Rethink America's Corn System." *Scientific American* March 5, 2013. https://www.scientificamerican.com/article/time-to-rethink-corn/. Accessed September 22 2017.

Gober, Patricia, Anthony J. Brazel, Ray Quay, Soe Myint, Susanne Grossman-Clarke, Adam Miller, and Steve Rossi. 2010. "Using Watered Landscapes to Manipulate Urban Heat Island Effects: How Much Water Will It Take to Cool Phoenix?" *Journal of the American Planning Association* 76(1): 101–121.

Gober, Patricia, Ariane Middel, Anthony Brazel, Soe Myint, Heejun Chang, Jiunn-Der Duh, and Lily House-Peters. 2012. "Tradeoffs between Water Conservation and Temperature Amelioration in Phoenix and Portland: Implications for urban sustainability." *Urban Geography* 33(7): 1030–1054.

Gober, Patricia, Ray Quay, and Kelli L. Larson. 2016. "Outdoor Water Use as an Adaptation Problem: Insights from North American Cities." *Water Resources Management* 30(3): 899–912.

Hamelinck, Carlo. 2013. "Biofuels and Food Security Risks and Opportunities." ECOFYS Research Paper. http://www.ecofys.com/files/files/ecofys-2013-biofuels-and-food-security.pdf. Accessed September 22 2017.

Harlan, S. L., A. J. Brazel, L. Prashad, W. L. Stefanov, and L. Larsen. 2006. "Neighborhood Microclimates and Vulnerability to Heat Stress." *Social Science and Medicine* 63(11): 2847–2863.

Hart, Melissa A., and David J. Sailor. 2009. "Quantifying the Influence of Land-use and Surface Characteristics on Spatial Variability in the Urban Heat Island." *Theoretical and Applied Climatology* 95: 397–406.

Health Canada. 2017. "Drinking Water Advisories: First Nations South of 60." https://www.canada.ca/en/health-canada/topics/health-environment/water-quality-health/drinking-water/advisories-first-nations-south-60.html. Accessed September 22 2017.

Hill, T., and C. Polsky. 2007. "Development and Drought in Suburbia: A Mixed Methods Rapid Assessment of Vulnerability to Drought in Rainy Massachusetts." *Global Environmental Change, Part B: Environmental Hazards* 7:291–301.

Hindustan Times. 2016. "Water Crisis: Bhopal Residents Facing a Trickle-down Effect." http://www.hindustantimes.com/bhopal/water-crisis-bhopal-residents-facing-a-trickle-down-effect/story-wd5cENfartjfYn1vqa576O.html. Accessed June 16 2017.

Hoekstra, Arjen Y. and Ashok K. Chapagain. 2007. "Water Footprints of Nations: Water Use by People as a Function of Their Consumption Pattern." In Integrated Assessment of Water resources and Global Change, *Water Resources Management* 21(1): 35–48.

House-Peters, Lily, Bethany Pratt, and Heejun Chang. 2010. "Effects of Urban Spatial Structure, Sociodemographics, and Climate on Residential Water Consumption in Hillsboro, Oregon." *Journal of the American Water Resources Association* 46(3): 461–472.

Kallis, George, and Harris Coccossis. 2003. "Managing Water for Athens: From the Hydraulic to the Rational Growth Paradigm." *European Planning Studies* 11(3): 245–261.

Klein, Gary, Martha Krebs, Valery Hall, Terry O'Brien, and B. B. Blevins. 2005. "California's Water-Energy Relationship." CEC-700-2005-011-SF. Sacramento: California Energy Commission. http://www.energy.ca.gov/2005publications/CEC-700-2005-011/CEC-700-2005-011-SF.PDF. Accessed September 16 2017.

Konar, M., C. Dalin, S. Suweis, N. Hanasaki, A. Rinaldo, and I. Rodriguez-Iturbe. 2011. "Water for Food: The Global Virtual Water Trade Network." *Water Resources Research* 47, W05520, https://doi.org/10.1029/2010WR010307.

Kruskal, Joshua. 2016. "Avocado Demand Fueling Deforestation, Unrest in Mexico." *International Policy Digest* September 7 2017. https://intpolicydigest.org/2016/09/07/avocado-demand-fueling-deforestation-unrest-mexico/. Accessed September 22 2017.

Kuckro, Rod. 2014. "Receding Lake Mead Poses Challenges to Hoover Dam's Power Output." https://eenews.net/stories/106002129. Accessed June 9 2017.

Lerman, Susannah, Victoria Turner, and Christopher Bang, 2012. "Homeowner Associations as a Vehicle for Promoting Native Urban Biodiversity." *Ecology and Society* 17(4): 45. https://doi.org/10.5751/ES-05175-170445. Accessed on June 15 2017.

Lewis, Lori. 2017. "Water in Crisis – Spotlight on the Asian Food Crisis." https://thewaterproject.org/water-crisis/water-in-crisis-asian-food-crisis. Accessed September 22 2017.

Malik R. P. S. 2002. "Water–Energy Nexus in Resource-poor Economies: The Indian Experience." *Water Resources Development* 18(1): 47–58.

Maupin, Molly A., Joan F. Kenny, Susan S. Hutson, John K. Lovelace, Nancy L. Barber, and Lindsey, Kristin. 2010. "Estimated Use of Water in the United States in 2010." United States Geological Survey Circular. https://pubs.usgs.gov/circ/1405/pdf/circ1405.pdf. Accessed June 8 2017.

Meyers, Stephan, Allison Williams, and Peter Chan. 2011. "Energy and Economic Impacts of U.S. Federal Energy and Water Conservation Standards Adopted From 1987 Through 2012." Berkeley, CA: Lawrence Berkeley Lab https://pubarchive.lbl.gov/islandora/object/ir%3A158812/datastream/PDF/view. Accessed June 14 2017.

National Oceanic and Atmospheric Administration (NOAA). 2017. "Gulf of Mexico "Dead Zone" is the Largest Ever Measured." http://www.noaa.gov/media-release/gulf-of-mexicodead- zone-is-largest-ever-measured. Accessed January 29 2018.

Oke, Timothy R. 1982. "The Energetic Basis of the Urban Heat Island." *Quarterly Journal of the Royal Meteorological Society* 108(455): 1–24.

Overturf, Michael. 2012. "How the Price of Power is Set." https://www.forbes.com/sites/energysource/2012/12/26/how-the-price-for-power-gets-set/#6dd97039702c. Accessed June 11 2017.

Partners for the Saskatchewan River Basin. 2009. "From the Mountains to the Sea: Summary of the State of the Saskatchewan River Basin." http://www.saskriverbasin.ca/public/uploads/flipbook/State_of_the_Basin_Report.pdf. Accessed September 26 2017.

Paterson, Willa, Richard Rushforth, Benjamin L. Ruddell, Megan Konar, Ikechukwu C. Ahams, Jorge Gironás, Ana Mijic, and Alfonso Mejia. 2015. "Water Footprint of Cities: A Review and Suggestions for Future Research." *Sustainability* 7: 8461–8490. https://doi.org/10.3390/su7078461.

Patrick, Robert J. 2011. "Uneven Access to Safe Drinking Water for First Nations in Canada: Connecting Health and Place Through Source Water Protection." *Health & Place* 17: 386–389.

Renewable Fuels Association. 2016. "Global Ethanol Production." https://www.afdc.energy.gov/data/10331. Accessed September 25 2017.

Renewal Fuels Association. 2011. "Fuel Ethanol Industry Guidelines, Specifications, and Procedures." http://www.ethanolrfa.org/wp-content/uploads/2015/10/Industry-Guidelines-Specifications-and-Procedures.pdf. Accessed September 23 2017.

Robbins, Paul. 2007. *Lawn People: How Grasses, Weeds, and Chemicals Make Us Who We Are*. Philadelphia: Temple University Press.

Rogers, Peter. 2009. "Water – Energy Nexus: Sustainable Urbanization in the Greater Mekong Subregion." http://scholar.google.ca/scholar?q=Water+-+Energy+Nexus:+sustainable+urbanization+in+the+Mekong+Basin&hl=en&as_sdt=0&as_vis=1&oi=scholart&sa=X&ved=0ahUKEwiQpOe02rbUAhVG0GMKHV_FBZ8QgQMIIzAA. Accessed June 11 2017.

Ruddell, Darren M., Sharon. L. Harlan, Susanne Grossman-Clarke, and Alexander. Buyanteyev, A. 2010. "Risk and Exposure to Extreme Heat in Microclimates of Phoenix, AZ." In *Geospatial Techniques in Urban Hazard and Disaster Analysis*, edited by P. Showalter and Y. Lu, 179–202. Dordrecht, Netherlands: Springer.

Stillwell, Ashlynn, Cary W. King, Michael E. Webber, Ian J. Duncan, and Amy Hardberger. 2010. "Energy-Water Nexus in Texas." *Ecology and Society* 16(1) https://www.ecologyandsociety.org/vol16/iss1/art2/. Accessed June 14 2017.

Times of India. 2017. "Bhopal: Parched Kolar Cries for Water Arrangement." http://timesofindia.indiatimes.com/city/bhopal/bhopal-parched-kolar-cries-for-water-arrangement/articleshow/58037021.cms. Accessed June 16 2017.

Unger, János. 2004. "Intra-urban Relationship Between Surface Geometry and Urban Heat Island: Review and New Approach." *Climate Research* 27(3): 253–264.

US Bureau of Reclamation. 2017. "Hoover Dam." https://www.usbr.gov/lc/hooverdam/history/storymain.html. Accessed June 14 2017.

US Congress. 2007. "Energy Independence and Security Act of 2007." Energy Independence and Security Act of 2007. Accessed September 26 2017.

US Department of Agriculture, Economic Research Service. 2017. "Background: Ethanol, Biodiesel, Biodiesel Co-products." https://www.ers.usda.gov/topics/farm-economy/bioenergy/background.aspx.g. Accessed on June 15 2016.

US Department of Energy, 2013. "US Energy Sector Vulnerabilities to Climate Change and Extreme Weather." https://energy.gov/downloads/us-energy-sector-vulnerabilities-climate-change-and-extreme-weather. Accessed September 23 2017.

US Department of Energy. 2014. "Energy-Water Nexus: Challenges and Opportunities." https://energy.gov/sites/prod/files/2014/07/f17/Water%20Energy%20Nexus%20Full%20Report%20July%202014.pdf. Accessed September 16 2017.

US Department of Energy. 2015. "The Water-Energy Nexus: Challenges and Opportunities." https://www.energy.gov/under-secretary-science-and-energy/downloads/water-energy-nexus-challenges-and-opportunities. Accessed June 14 2017.

US Environmental Protection Agency. 2013. "Outdoor Water Use." EPA-832-F-06-005 Accessed on June 15, 2017 https://www.epa.gov/sites/production/files/2017-03/documents/ws-factsheet-outdoor-water-use-in-the-us.pdf.

Walsh, Breandan P., Seán N. Murray, and D. T. J. O'Sullivan. 2015. "The Water Energy Nexus, an ISO50001 Water Case Study and the Need for a Water Footprint Network. 2017." Product Water Footprints Coffee and Tea. http://temp.waterfootprint.org/?page=files/CoffeeTea. Accessed August 25 2017.

Water Footprint Network. 2017. "Water Footprint of Crop and Animal Products: A Comparison." http://waterfootprint.org/en/water-footprint/product-water-footprint/water-footprint-crop-and-animal-products/. Accessed June 8 2017.

Zhuo, La, Mesfin M. Mekonnen, Arjen Y. Hoekstra, and Yoshihide Wada. "Consumptive Water. Footprint and Virtual Water trade Scenarios for China—With a Focus on Crop Production, Consumption and Trade." *Environment International* 94(2016): 211–223.

5

Meaning, Purpose, and Value of Water

Out of Water: The Distant Alarm Comes Closer
"Water, unlike religion and ideology, has the power to move millions of people. Since the very birth of human civilization, people have moved to settle close to water. People move when there is too little of it; people move when there is too much of it. People move on it. People write and sing and dance and dream about it. People fight over it. And everybody, everywhere and every day, needs it. We need water for drinking, for cooking, for washing, for food, for industry, for energy, for transport, for rituals, for fun, for life. And it is not only we humans who need it; all life is dependent upon water for its very survival."
(Mikhail Gorbachev, Civilization: The Magazine of the Library of Congress, Oct/Nov 2000).

5.1 Introduction

In March 2017, a court in the Northern Indian State of Uttarakhand granted the Ganges River and its tributary, the Yamuna, legal status as living human entities and appointed custodians to conserve and protect them (Safi 2017). Water is not an economic good in the usual sense. Its

P. Gober, *Building Resilience for Uncertain Water Futures*,
https://doi.org/10.1007/978-3-319-71234-5_5

unique properties (e.g., flow resource, lack of substitutes, essential for life, spiritual and aesthetic dimensions, impacts on communities) lead to widespread agreement that it should be governed by societal institutions at a range of scales from local and regional to national and international levels (United Nations 2009). Societies organize the use of water based on a set of rules. The development and implementation of these rules is a value-based exercise and political process. The rational optimization method allocates water to competing users, making assumptions about human values (e.g., that water decisions should be rational, efficient, and based on scientific evidence). These assumptions may or may not be true. Underlying values about water are beliefs about economic efficiency, social equity, and environmental protection, the need to provide for future generations, aesthetic and spiritual concerns, the role of government in civil society, and the human right to water. The decision about how to manage a reservoir, for example, is a complex interplay among needs for agriculture, industry, urban development, hydropower generation, flood control, ecosystem health, and recreation. Today's water policy decisions also evolve from attitudes about climate change, belief in the efficacy of climate science, and competing values among users. They reflect deep-seated beliefs about the rights and responsibilities of individuals and groups in society and the role of science in decision-making. Effective water governance acknowledges competing value systems and establishes decision-making institutions that adjudicate disagreements about the meaning, purpose, and value of water.

5.2 Climate Change Attitudes

Public attitudes about climate change background many current discussions about water problems. Climate change attitudes reflect profound differences across and within nations about the efficacy of climate science, the relationship between humans and the environment, and the legitimate role of government in resource management. In a recent survey of climate change attitudes using public opinion research, the Pew Research Center found that a solid majority of respondents from the 40 countries studied believed that climate change was a serious problem and

78% supported the idea of having their country reduce CHG emissions in line with the international agreement negotiated in Paris. Despite this widespread global concern about the seriousness of climate change, there are significant geographic differences in perceived impacts on individuals and countries. Respondents from the US and China, the countries that are most responsible for GHG emissions, were in 2015 the least concerned about the problem, less likely to believe that climate change is affecting people now, and less likely to believe that climate change will harm them personally. Only 15% of Chinese and 30% of Americans agreed with the statement "climate change will harm me personally" compared to 63% in Latin America and 61% in Africa. In most countries, publics believe that policy to reduce GHG emissions should come disproportionately from the countries that are primarily responsible for the problem including China, the US, the European Union, and India (Stokes et al. 2015).

There is an emerging global consensus that technology and policy alone cannot solve potential problems of climate impacts, and lifestyle changes will be required. Even in the US where there is strong belief in the capacity of technology to solve environmental problems, 66% of respondents to a Pew Research survey believed that significant lifestyle changes are required to address climate change (Stokes et al. 2015).

Divergent public attitudes about the *science* of climate change are problematic for those who seek to link policy and decision-making to evidence-based science. Belief in climate science and trust in climate scientists in the US are highly polarized and strongly related to political party affiliation (Funk and Kennedy 2016). Liberal Democrats are far more likely than conservative Republicans to trust scientists to provide understanding of the causes of climate change, believe that scientists understand the causes of climate change very well, perceive consensus among scientists about the causes of climate change, and believe that climate research reflects the best available evidence most of the time. In contrast, conservative Republicans are far more likely to believe that climate research findings reflect scientists' desire to promote their career advancement and their political leanings. These results signal deep and enduring skepticism about the motivations of scientists and the veracity of their climate change assessments. Just 67% of survey respondents in

2017 thought that scientists should have a major role in the policy-making process despite the fact that downscaled global climate models are often the first step in regional and local water resource assessments.

These attitudes inform both the highly politicized public debate about climate change in the US (Marcin 2017) and the evolving role of science in public debate and environmental policy. Embedded in IPCC assessments is the view that scientific facts are or should be the basis upon which we formulate public policy. In fact, scientific knowledge is dependent on the political context in which it occurs. It is possible for people with a variety of interests and values to call upon different bodies of knowledge to help them understand and respond to a particular problem (Sarewitz 2004). Knowledge relevant to one way of understanding the problem may be different from another perspective. Disagreements about the impacts of climate science and about the motivations and trustworthiness of climate scientists reflect, in part, these deep-seated cultural values and political perspectives.

Many environmental disputes are not resolved with scientific evidence alone. Social goals and political processes are at the core of environmental policy in democratic societies. Science may be not so relevant in reducing uncertainty about future climate and hydrological conditions (Chap. 3) but useful in developing future scenarios that respond to differing sets of values, different ways of understanding the problem, and different policy preferences (Chap. 8).

5.3 Is Water a Common Good or a Commodity?

Water governance embodies the tension between competing views of water as a common good and commodity. The commons view of water considers its status as a flow resource, essential for life and ecosystem health, tightly bound to communities, and with cultural and spiritual dimensions that reflect place-based practices (Bakker 2007). Theoretically, collective management of water as a common good prevents market and state failures, reflects the deep cultural and aesthetic meanings associated with water and water use, and protects public health. Nineteenth-century

European and North American cities took control of water resources to protect public health from outbreaks caused by contaminated water from uncontrolled taps. The view of water as a common good dominated water resource management for most of the twentieth century. Even today, the European Association of Public Operators asserts that water is a "common good." Policies stemming from this principle "should ensure the long-term protection of water resources and infrastructure to access them, and that revenues generated from water services should be allocated and invested for the benefit of all society" (European Association of Public Operators 2017).

Twenty-five years ago, market environmentalists fostered the view that environmental problems are in fact market failures in the sense that societal institutions do not adequately protect individual property rights or efficiently allocate resources (Anderson and Leal 2001). In this context, water is an economic good that is no different from other essential goods such as food and utility services. Theoretically, private companies are responsive to their customers, and appropriate pricing encourages conservation and reduces wasteful practices as water prices rise with increasing scarcity. Proponents of market environmentalism asserted that environmental externalities are included in price, and market mechanisms both promote the most efficient use of the resource and reduce environmental harm (Finger and Allouche 2002; Johnstone and Wood 2001). This view had support from the international water establishment as Guiding Principle No. 4 of the 1992 International Conference on Water and the Environment in Dublin, Ireland, declared that water is an economic good albeit with some need for public oversight.

> Principle No. 4 – Water has an economic value in all its competing uses and should be recognized as an economic good. Within this principle, it is vital to recognize first the basic right of all human beings to have access to clean water and sanitation at an affordable price. Past failure to recognize the economic value of water has led to wasteful and environmentally damaging uses of the resource. Managing water as an economic good is an important way of achieving efficient and equitable use, and of encouraging conservation and protection of water resources. (International Conference on Water and the Environment 1992)

The dichotomy between water as a commodity and common good is overly simplistic. In some places, a high degree of state involvement co-exists with private management. In the UK, for example, the state heavily regulates private water companies with price controls and quality standards. In the Netherlands and France, the state also retains ownership while privatizing water services (Bakker 2007). The distinction between water as a commodity and common good also fails in many developing world settings where public water companies provide inadequate services to many areas, causing many to rely on informal, for-profit providers for their water supply needs, as we have seen in the rise in private tankers to deliver water in some megacities.

Thus, water provision often mixes commons and commody concepts to reflect place-based conditions and value systems. Governments can retain central control of the water allocation process, given water's growing scarcity and centrality to human life, but there is potential to achieve greater efficiency by privatizing some functions of water management. The California Water Bank, for example, buys water rights for temporary and long-term water leases from private holders, usually farmers, at an administratively set price. The Bank then sells the rights to other users, including other farmers growing higher-value crops, cities, or environmental programs, also at an administratively set price. The water market now accounts for approximately 5% water use in the state with about two million acre-feet (2.5 billion m^3) changing hands annually. The market evolved from one dominated by short-term contracts to permanent trades that have reordered the pattern of water use in favor of farmers growing high-value crops, cities, and environmental protection. Between 2003 and 2011, water purchases for the environment represented 20% of commitments in the water market (Hanak and Stryjewski 2012).

The Colorado-Big Thompson (CBT) Project, Northern Colorado's largest water provider, is a federal water system overseen by a locally appointed board. It brings water from the headwaters of the Colorado River in the Rocky Mountains through the Adams Tunnel to the Northern Colorado Water Conservancy District (Fig. 5.1). The District allows users to trade water rights. Built originally in the 1930s to serve the region's irrigated agricultural economy, the system now provides water to

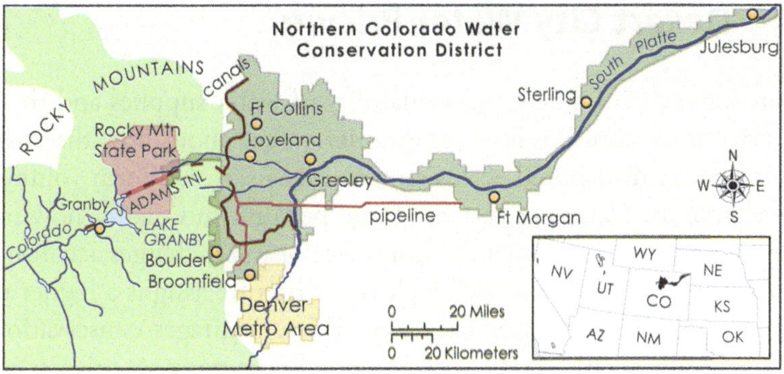

Fig. 5.1 Colorado-Big Thompson Project

rapidly growing cities and the oil and gas industry. Industry and municipalities own a majority of the water rights, and the District leases water rights to farmers on a temporary annual basis. The District sets the terms of the transactions and has final approval of all transactions. Prices are determined at auctions. CBT prices tripled between 2010 and 2015. They are a barometer of economic value in a trans-boundary water system with a streamlined auction process (West Water Research 2015). Higher water prices stem, in part, from growth in the region's fracking industry that is the highest bidder for water on the open market. Also significant is demand from the home building industry where the cost of water rights factors into new water hook-ups. The water market has moved the resource from agriculture to higher-value uses, signaled scarcity in the form of rapidly increasing prices, and shifted the burden of acquiring new water rights to new homeowners.

Market-based systems have both strengths and weaknesses. Strengths are the potential to reduce ineffective and inefficient government management and motivate water transfer from lower- to higher-value uses. Weaknesses include the failure of the market-based model to adequately account for quality-of-life and social issues, the highly contested nature of allocation equity for a shared resource, lack of transparency, unintended consequences in complex environmental systems, and the potential for monopoly (Draper 2008, p. 498).

5.4 Desert City Water Pricing

Urban water prices reflect the availability of water supplies and the cost of service provision, but also community values about how the costs of operations are distributed across community users. Uniform volumetric rates charge for water use at a fixed rate per unit, no matter how much water is used. Block rates set the unit price of water charges according to the level of use. An increasing block rate structure assigns a higher price on a per unit basis as usage increases. This encourages conservation. A decreasing block rate charges customers a lower per unit price as water use increases—the more you use the cheaper it is. Decreasing block rates penalize customers who consume little water and discourage conservation. They are declining in popularity over time.

The desert cities of Phoenix and Tucson in Central Arizona use different rate structures to price water, reflecting differing levels of water scarcity and community values about water conservation, fairness, and equity. Phoenix has a volumetric system of water use charges, while Tucson has steeply increasing block rates (Fig. 5.2). Phoenix has always had a larger and more diverse water portfolio than Tucson. It developed as an oasis rather than as a desert city (Fig. 5.3). Early settlers created landscapes and

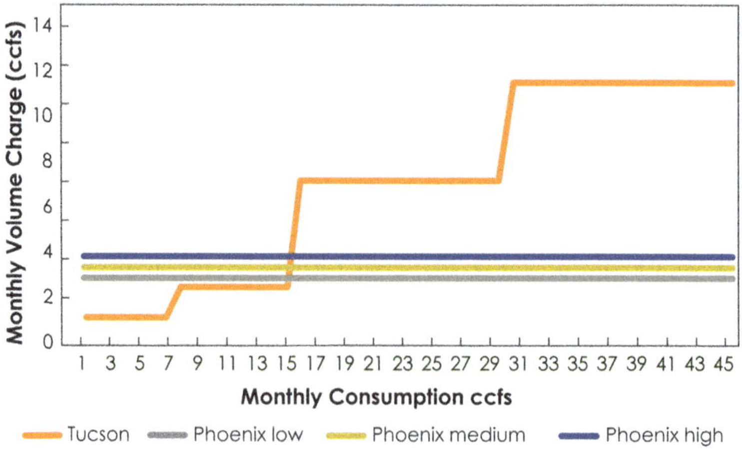

Fig. 5.2 Water rates in Phoenix and Tucson

Fig. 5.3 Phoenix is a desert city with oasis landscapes

gardens befitting the more humid climates from which they came, such as the Midwestern or Northeastern US (Gober 2006). Tucson, on the other hand, was more limited in its water availability. Residents developed a desert-oriented architecture, landscape, and urban design based on native trees and vegetation, fewer swimming pools, and less outdoor water use (Fig. 5.4). Tucson today uses an increasing block rate structure to assign rapidly increasing rates based upon the amount used. The rate structure begins, however, at a very low level, making basic human needs for indoor water relatively inexpensive. A majority of households do not surpass the second tier of the rate structure. Phoenix has a uniform rate structure that shifts slightly upward in the summer (June through September) and shoulder (April, May, October, and November) seasons to reflect greater demand for outdoor water. It does not price for conservation. Water was more plentiful in Phoenix and the local culture favored the view that water is a service for which individual households should pay. The average summer monthly usage charge for 30 ccf (1 ccf = 100 ft^3 = 748 gallons = 2.83 m^3) in 2016 was $80.60 in Phoenix compared

Fig. 5.4 Tucson is a desert city

to $146.05 in Tucson. The gallons of residential use per capita per day were 106 in Phoenix in 2013 and 82 in Tucson in 2016 (Tucson Water 2017, City of Phoenix Water Services Department 2014).

5.5 Prairie Potholes

Management of agricultural drainage in Canada's Prairie Provinces pits the interests of individual farmers against societal goals of flood control, water storage, and environmental protection. Ongoing debates about drainage reveal competing values about water, wetlands, and government regulation.

Prairie potholes are shallow depressions that lack perennial surface water connections. They are remnants of the last glacial age and ubiquitous in the Smith Creek Prairie region of North America (Figs 5.5 and 5.6). They are home to more than 50% of North American migratory waterfowl, and many species are dependent upon them for breeding and feeding. Potholes absorb surges of rain, snow, and floodwaters, thus reducing the

Fig. 5.5 Smith Creek agricultural drainage (natural) 2015. Ducks Unlimited Canada

Fig. 5.6 Smith Creek agricultural drainage (drained) 2015. Ducks Unlimited Canada

risk and severity of downstream flooding. They provide a hedge against drought as more surface water is available to support wildlife during dry years.

Since European settlement in the early twentieth century, Prairie farmers have drained more than 50% of the region's natural potholes to increase the size of their planted areas and make them easier to manage. Severe downstream flooding in recent years prompted questions about the role of drainage vis-à-vis land use change and climate change in intensifying flood events in river basins (Dumanski et al. 2015). Today's drainage discussion reflects differing viewpoints of land and water rights, responsibilities of landowners relative to their neighbors, and the legitimate role of government in regulating conflicts within its jurisdiction. The story of Prairie potholes shows how values dominate the public debate about changing hydrology, land management, and environmental protection in Western Canada.

The Prairie climate is semi-arid and plagued by significant variability in annual precipitation. This variability initially postponed development of the Canadian Prairies. John Palliser, leader of the 1857–1859 expedition by the Royal Geographical Society to the Canadian West, reported that the area was too dry for agricultural development. Later government officials saw the benefits of rich soil, thought it suitable for wheat farming, and advertised it to new immigrants, many from Eastern Europe, beginning in the early twentieth century. By 1925, the province of Saskatchewan produced over half of the wheat in the Dominion of Canada. Although the Dust Bowl years of the 1930s hit the region particularly hard, with many farmers losing their land, modern farming practices renewed the agricultural sector and rejuvenated the economy after World War II. Today, 39% of Canada's farmland is in Saskatchewan, and 32% is in neighboring Alberta (Statistics Canada 2016a). In 2016 the Prairie Provinces of Saskatchewan, Alberta, and Manitoba produced 95% of Canada's barley, 99% of its grain, 93% of its corn, 100% of its lentils, and 97% of its wheat (Statistics Canada 2016b).

European settlement of the region involved continuous efforts to alter the landscape to make it more hospitable for large-scale agricultural development. The drainage of agricultural lands allowed farmers to gain earlier access to their fields, reduced the nuisance of farming around potholes, and increased the acreage planted—activities that have significant benefit to the farmer specifically and to agricultural production regionally (Saskatchewan Water Security Agency 2011). In the Smith Creek Subbasin of the Saskatchewan River Basin, the percentage devoted to wetlands declined from 96 square kilometers (24% of the basin area) in 1958 to 40.3 square kilometers (10% of the basin area) in 2009 (Dumanski et al. 2015). However, benefits to upstream farmers came at a cost to neighboring farmers, migratory birds, lost habitat, and downstream residents via increased flooding and diminished water quality as nutrients from agricultural production are flushed downstream rather than processed in adjacent wetlands.

Scientific evidence links agricultural drainage to increasing flood problems and raises the knotty issue of potential climate change impacts on this rapidly evolving environmental system. Drainage projects open links between self-contained (so-called non-contributing) areas to stream flow in local and regional rivers and streams. Simulations show that complete drainage of existing wetlands would have increased the peak of an already disastrous 2011 flood by 78% and the yearly volume of stream flow by 32% (Pomeroy et al. 2015). Landscape modification allows little or no residual local storage, and unregulated drainage ditches transport water from one local wetland to another, causing flood damage to adjacent croplands and communities surrounding the terminal wetland. Ducks Unlimited, a non-profit conservation organization devoted to the preservation of wetlands, called for more regulation of agricultural drainage, ranking the Prairie Pothole Region number 1 in its designation of the 25 most important and threatened waterfowl habitats in North America.

The problem of Prairie agricultural drainage pits the interests of individual farmers and collective gains to the regional economy against the social benefits of water storage, flood attenuation, wildlife habitat, and water quality. In 2013, the Province of Saskatchewan's Water Security Agency issued a 25-year plan and recently moved to license and regulate agricultural drainage (Saskatchewan Water Security Agency 2013, 2017). Current policy requires licensing all drainage works retroactively. Landowners must obtain permission to drain from downstream landowners, and some license holders will need to install culverts and gates, among a larger suite of other potential regulations (Saskatchewan Water Security Agency 2015).

Public discussion reflects different viewpoints about who should pay the costs associated with managing drainage. The Saskatchewan Farm Stewardship Association asserts that it is impossible to register and retroactively license all drainage works in the province. Projects completed before a certain date need to be grandfathered into a new licensing system. It is unfair, they say, to ask farmers to pay to ameliorate flood drainage problems that extend beyond their agricultural lands and include natural habitat areas and naturally drained lands adjacent to their properties. In addition, they contend that the social benefits of mitigating flood hazard extend far beyond the agricultural sector. Therefore, society more generally should absorb the mitigation costs of agricultural drainage. In a recent article entitled "Farming for Ducks," farmers objected to land regulation.

> As urbanization and wildlife protection issues gain greater traction in public opinion, special interest groups are poised to hijack the public agenda and diminish the rights of agricultural producers to have control over their land. To ensure that future generations can prosper, the Sack FSA exists in part to ensure that control of agricultural land uses stays where it belongs—in the hands of producers. (Saskatchewan Farm Stewardship Association 2017)

Ducks Unlimited counters with the perspective that "wetlands are viewed by some people as an inconvenience. That's why many have been drained for industrial and agricultural development" (Ducks Unlimited 2017).

This is a story of competing values about land rights, flood risk, economic growth, environmental services, public relations, and social responsibilities in a democratic society. Essential to effective management of drainage are governance institutions and policies capable of achieving outcomes that are both socially equitable and environmentally sustainable from the perspective of multiple stakeholders. Any viable resolution needs to account for viewpoints of farmers, environmentalists, scientists, government regulators, industry, and municipal water managers. In

February 2017, the Saskatchewan Water Security Agency approved a new agreement, involving 73 landowners of more than 18,000 acres who agreed to establish a drainage network involving 113 quarter-sections. Landowners control the project and agreed to drainage works and controlled releases for existing drainage that protect downstream farmers from increased flooding. The project also restores 34 acres of wetlands on existing drainage and 21 acres of wetland retention on new drainage (Saskatchewan Water Security Agency 2017).

5.6 Indigenous Water Rights

Indigenous peoples worldwide have seen traditional water sources depleted for the benefit of dominant, non-native society. Depletion and contamination have made it difficult for native peoples to practice long-standing, water-dependent occupations such as farming, hunting, and fishing. Also lost are the spiritual, ceremonial, and communal aspects of water. National governments developed water laws to protect the rights of European settlers and transnational corporations at the expense of the cultures and economies of Indigenous peoples (Getches 2005). Concern about Indigenous land and water rights in Australia, New Zealand, the Americas, and Europe inspired a body of academic literature and a political movement that call for the reassertion of Indigenous water rights (Battiste 2000; Turner 2006, Coulthard 2008). Hidalgo et al. (2017, p. 68) described the loss of water security to native peoples in the Andean Highlands of Latin America where a traditional culture based on collective institutions faced powerful actors, national elites, and transnational enterprises that favored reallocations of water rights to "supposedly more productive more efficient uses." New market prices for water and rules that favor individual over collective rights were the basis for Chile's economic development in the 1980s. More recently, similar relationships have benefited flower growers in Ecuador at the expense of Indigenous small farmers.

Ethnographic studies document the deep knowledge of water distribution that was required to harvest fish and wildlife over thousands of years before Anglo-European settlement of the Americas and elsewhere (Keen

2004; Bandler 1995). Indigenous peoples developed complex systems of water resource management to deal with variations in rainfall and temperature. Social conventions regulated access. Rules varied with the abundance and reliability of the resource (Hiscock 2008). In Central Arizona, the prehistoric Hohokam (a Pima Indian word meaning "those who have disappeared") developed a large and complex civilization based on irrigated agriculture. At its peak in 1000 AD, the civilization numbered 40,000, supported by an estimated 110,000 acres under cultivation. A high level of regional cooperation would have been required to harness and deliver water to a population on that scale under arid climate conditions (Abbott 2000). Beyond its value in productive livelihoods, water had meaning to people through social institutions, and water linked to local customs, norms, and rituals. Indigenous conceptualizations of water extended far beyond the notion of cultural heritage to include sociality, sacredness, identity, and life-giving (Jackson 2006). In Latin America, political struggles for self-determination reveal the role of water in Andean culture and its importance for a sense of belonging, identity, mutual rights and responsibilities, and exclusiveness (Hidalgo et al. 2017). The Maoris in New Zealand claim a special relationship with water based on the historical use of rivers for drinking water, spiritual beliefs, and fishing and shellfish harvesting (*The Economist* 2017).

The global movement for Indigenous self-determination led to the 2007 United Nations Declaration on the Rights of Indigenous Peoples (with votes against by Australia, New Zealand, Canada, and the US). This Declaration reaffirmed the rights of Indigenous peoples to development based on their own needs and interests. Included in the resolution was the "urgent need to respect and promote the inherent rights of indigenous peoples which derive from their political, economic and social structures and from their cultures, spiritual traditions, histories and philosophies, especially their rights to their lands, territories and resources" (United Nations Declaration on the Rights of Indigenous Peoples 2007). Embedded in the resolution is codification of the rights of Indigenous peoples to adhere to their own values as self-determining nations with rights to culture, identity, language, health, education, and traditions.

Indigenous peoples face daunting challenges in translating the 2007 UN Declaration into on-the-ground water rights, often because their communal values conflict with neo-liberal claims of efficiency, rationality, objectivity, individualism, and optimization. Also significant are the invisible ways in which water factors into identity and lifestyles. In Australia, the concept of "cultural flows" captures Aboriginal-specific needs, but the concept has led to very few reallocations that improve the lives of people on the ground. Neo-liberal reforms that convert water into tradable property rights generally assume that water is a private rather than communal resource as seen by most Indigenous peoples (Jackson et al. 2012). In Canada, Indigenous peoples have made modest gains in re-negotiating traditional water rights. Focus is on managing problems of jurisdictional authority rather than on dealing with larger claims of self-determination and reallocation. Regional forums for water policy reform typically treat First Nations communities as one of many stakeholders rather than as self-determined nations, exercising legitimate historical claims to water rights (von der Porten and de Loë 2014a, b).

In the US, federal law stipulates that Native American water rights have a priority date and quantity. Because they date from as early as the time of reservation establishment (usually in the late nineteenth and early twentieth centuries), Indian water rights are often the most senior rights in a particular river basin. The legal system has resolved a number of these cases and overseen negotiated settlements in others. Critical to these cases is the Winters Doctrine of 1908 in which the Supreme Court of the US clarified water rights for Indian reservations. In 1888, the federal government designated the Fort Belknap Indian Reservation in the State of Montana along the Milk River from a much larger area set aside for tribes. The agreement was silent on the issue of water rights for the reservation. Subsequently, settlers occupied the area, built dams and reservoirs, and prevented water use for agricultural purposes by the reservation. The settlers claimed rights to the water based on the first-in-time-first-in-right (FITFIR) principle of prior appropriation; they claimed to have appropriated the water for beneficial use before the Native Americans on the reservation. Native Americans, in turn, claimed their water rights were implicit in the 1888 creation of the Milk River Reservation. In the 1908 case of *Winters v. the United States*, the Supreme Court ruled that the government created reservations to support self-sufficient agricultural communities, and thus water rights are implicit in that agreement (US Supreme Court 1908). In a series of cases over the next 100 years, the

Supreme Court upheld Indian water rights, the priority of federal over state courts, and the ability of Indians to sell their water rights to new owners.

The federal government was less than diligent in translating case law into water rights to oversee economic development on Indian lands, although some progress occurred to settle disputes recently. Four long-standing disputes were resolved in December 2016, and President Obama signed them into law in the closing days of his administration in January 2017. Legislation authorized $422 million in funding to the Blackfeet Tribe of Montana to provide clean drinking water and other water-related infrastructure projects and $28 million for the Pechanga Band of Luiseño Mission Indians, located in Southern California, to gain secure water supplies. The legislation also secured for the Choctaw and Chickasaw Nations of Oklahoma the right to use and benefit from water resources within their historic treaty territories, and effected a settlement enacted in 1988 for five San Luis Rey Bands in Southern California that did not require additional funding.

5.6.1 Cumberland House

Upstream dam development inexorably changed the lives of Indigenous peoples in Cumberland House at the Delta of the Saskatchewan River in the Boreal Region of Northern Canada (Fig. 5.7). The region had a long history of pre-European settlement and served as a travel corridor and gathering place for Aboriginal peoples. The Saskatchewan River was a major west–east corridor linking the Canadian Rockies to the Atlantic Ocean (via Hudson Bay). European explorers and fur traders traveled the river during the eighteenth and nineteenth centuries. Cumberland House, established in 1774, is the oldest continuously occupied settlement in Western Canada. It was a crossroads for fur-trading routes developed by the British-based Hudson's Bay Company and the Montreal-based North West Company.

Cumberland House today consists of two communities: the Northern Village of Cumberland House and the Cumberland House Cree Nation. They are of Metis and Cree ancestry and together numbered 1466 resi-

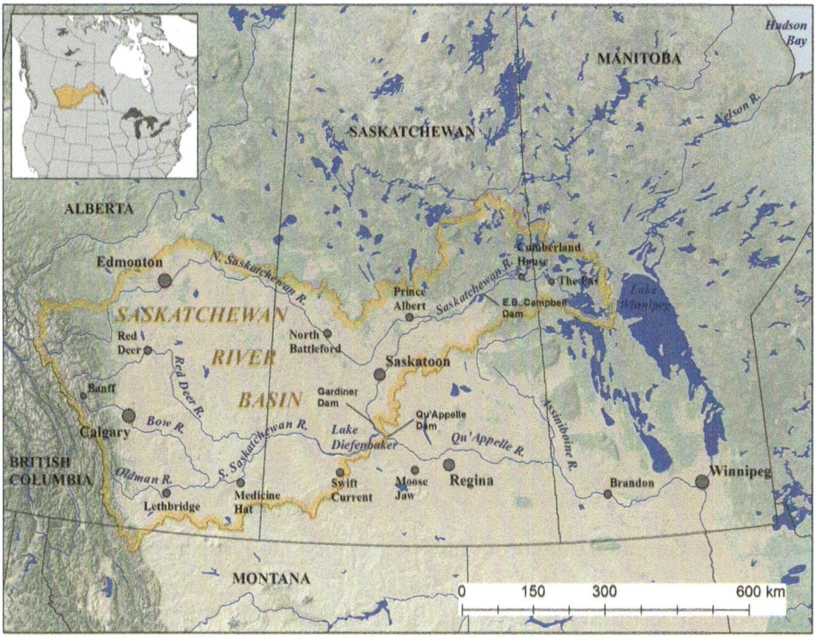

Fig. 5.7 Cumberland House. Chris DeBeer/Global Institute for Water Security/ University of Saskatchewan

dents in 2016 (Statistics Canada 2017). Construction of the E.B. Campbell dam upstream from Cumberland House altered traditional livelihoods of hunting, fishing, trapping, and serving as travel guides (Fig. 5.8). The Saskatchewan Power Company (SaskPower), a Crown Utility, led hydropower development of the lower Saskatchewan River Basin in the 1960s to facilitate economic development and provide clean energy to meet peak demand. The Province of Saskatchewan granted the dam a 50-year license in 1985 applied retroactively to 1966. The Province also built two other upstream dams (Gardiner Dam in 1966 and Francois Finlay dam in 1985), but concern is with E.B. Campbell because of its proximity to Cumberland House and its "hydropeaking" capacity (Andrews 2015). During early years of development, SaskPower completely shut off water flows when there was no power, creating disruption of downstream flows with dire consequences for fishing and trapping. Subsequently, SaskPower

Fig. 5.8 E.B Campbell Dam upstream from Cumberland House

negotiated a settlement of $20 million with North Village Cumberland House (NVCH) to compensate it for environmental losses associated with dam operations.

More than 50 years of scientific evidence, combined with personal testimonials, now provide a stark and multi-dimensional picture of the impacts of dam construction for downstream Indigenous residents. The marshes surrounding Cumberland House form the largest inland river delta in North America. It is biologically diverse; more than 80% is wetland and home to migrating bird populations. Dam operations reduced flows, starved the delta ecosystems of nutrient-rich sediments, and changed the historic flow regime to which plants and animals had adapted over thousands of years (Hassanzadeh et al. 2017). A natural river in this part of the world relies on snowmelt from the Rocky Mountains and rainfall over the Prairies and has its highest flows during the spring and summer. A dam designed for hydropower stores summer water in its reservoir, and then runs it through turbines in winter months when electrical demand is highest.

The result is a fundamentally altered ecosystem where plants and animals struggle to maintain natural connectivity between lakes and wetlands, thus affecting the integrity of freshwater ecosystems, limiting access to spawning and nursery areas for fish, reducing riparian habitat, and damping sediment transport and deposition (Hassanzadeh et al. 2017). High water in late spring provides rearing habitat for spring-spawned fish such as sturgeon, northern pike, and walleye (Paul 2013). Local residents note the decline in the moose population and waterfowl, earlier departure of geese and ducks, increased presence of non-native birds, and more dead fish (Abu et al. 2017). Ecologists describe profound disruptions in the local ecosystem, and geomorphologists note that the main channel of the Saskatchewan River is wider and deeper; there are fewer smaller streams; and the inundation patterns during flood years (i.e., floodplain maps) are radically changing (Smith et al. 2014).

Local residents provide compelling evidence of what dam-induced environmental change means for their lives and livelihoods. People who depend upon fishing, hunting, and trapping are no longer able to pursue these occupations and traditional activities. Equally significant, they are unable to pass these traditions from one generation to the next. Thousands of years of family heritage stops here. Andrews (2015) conducted in-depth qualitative surveys with local residents and described "invisible losses" for hunters, trappers, and fishers. They included disruptions to navigation (significant for someone who uses a boat to get around or uses snowmobiles and dogsleds for trapping in the winter) and the jarring sight of dead moose, deer, and rabbits unable to survive large fluctuations in water levels. Also significant is the end of cultural traditions like trapping and trading pelts, the inability to pass valued knowledge about the ecosystem and fishing from one generation to the next, and loss of identity stemming from their connection with the delta environment.

Economic development of upstream communities and cheap, clean hydropower come at the expense of traditional livelihoods for First Nations communities. While efforts are under way to adjust flows to reduce the deleterious consequences of hydropeaking, it is unlikely that the natural ecosystems that supported a distinct culture and livelihoods of the Indigenous Cree and Metis populations and North America's largest bird sanctuary will return to pre-dam conditions. Downstream wild bird habitat, Indigenous water rights, and the glory of thousands of sea birds in the high grasses at sunset come at the expense of economic development, food production, and clean energy upstream.

5.6.2 People of the River: The Pima and Maricopa of Central Arizona

A second example of the clash in values about the water rights of Indigenous peoples involves the Gila River Indian Community's (GRIC) longstanding dispute with the US federal government over water rights. The dispute was resolved when President George W. Bush signed into law the Arizona Settlements Act on December 4, 2004. The agreement restored the community's rights to 653,500 acre-feet (more than 80 million m^3) of water or 23% of Arizona's contractual rights to Central Arizona Project (Colorado River) water (Bark and Jacobs 2009). This story illustrates GRIC's long journey to exercise its legal claims to water and its emerging role as a power broker in the Phoenix region.

The Pima (Akimel-O'odham) and Maricopa (Pee-Posh) Indians are thought to be the descendants of the prehistoric Hohokam civilization that occupied riparian lowlands in Central Arizona between 0 and 1400 AD. The Hohokam developed a complex and extensive canal system over hundreds of years of irrigating the desert. They used a combined system of primary, secondary, and tertiary canals and secondary laterals for irrigation purposes. At the peak of the civilization in 1000 AD, the population totaled 40,000, and 110,000 acres were in agricultural production. The civilization largely vanished by 1400. The Pima and Maricopa (People of the River) are probably the modern descendants of the prehistoric Hohokam.

The US government established GRIC along the Gila River in 1859 for use by the Pima and Maricopa. The reservation was located along the Gila River because of the importance of water to their way of life. Mexican and Mormon settlers arrived upstream after the Civil War and began diverting water for agricultural operations, depriving the Pima and Maricopa of access to the waters of the Gila River. The first reported shortage of water for Indian farmers occurred in 1868, but Indian agents did little but report the shortage to Washington as the federal government was more interested in stimulating settlement of the non-Indian population at that time (Navajo 2012). Some of the Pima and Maricopa moved to the adjacent Salt River Reservation, but the remaining people experienced famine conditions and ecological collapse, as the riparian habitat that supported fish and beaver disappeared (Fig. 5.9). The Pima and Maricopa cut down mesquite trees to buy food from non-Indian settlers upstream. Early explorers characterized the area as a lush green valley filled with mesquite thickets, cottonwood trees, and irrigated fields. When deprived of water from the Gila, it became dry and desiccated landscape unfit for human habitation. Thousands of Pima and Maricopa died of malnutrition during the "Starving Years" between 1895 and 1905.

Geographically, the early Native American settlement was set apart from early American and Mexican settlements. More recently, the growth of urbanized Phoenix spread to the borders of the reservation, and it now includes casinos, a resort, a spa, equestrian center, shopping centers, golf courses, and small airport.

Fig. 5.9 Pima Indian farmers building a dam for irrigation ca 1900. University of Southern California Libraries and California Historical Society

The tribe worked assiduously to establish its legal rights to water for more than 100 years, asserting that the Winters Doctrine gave it legitimate rights to water from the Gila River based on the date of the reservation's establishment (1859) and on the Practical Irrigated Acreage (the irrigation application rate times the amount of irrigable land on the reservation). The tribe argued that it was entitled to the entire flow of the Gila River based on the contention that the reservation would have expanded its agricultural production had it not been cut off from its water source by neighboring developments. Early court rulings sought to split river flows between GRIC and non-Indian neighbors, but the tribe objected, and decades of litigation followed without resolution. The US Congress authorized the Central Arizona Project (CAP) in 1968 to bring Colorado River water to Central Arizona with an eye toward setting aside a portion of this water to settle GRIC's water claims. In 1976, the Secretary of the Interior offered GRIC 173,100 acre-feet of Colorado River water. The tribe refused and continued pursuing its entitlements on the Gila River, asking for more than one million acre-feet (2.2 billion m^3) (DeJong 2014).

After several more decades of fruitless litigation and settlement offers, the tribe agreed in the 1980s to a negotiated settlement. In 1992, they approved a settlement framework. Twelve years later, the tribe agreed to the Arizona Water Settlement Act, restoring annual water rights of 653,500 acre-feet to the community and providing $200 million to rehabilitate its water infrastructure (DeJong 2007, 2014). For perspective, this water budget amounts to 23% of Arizona's rights to Colorado River water via the Central Arizona Project (Bark and Jacobs 2009). It is by far the largest potential source of water in the state to support either the renewal of agricultural production on the reservation or continued urban development, or some combination of the two. The tribe has been reticent to reveal its long-term plans beyond expressions of its intentions to reinvigorate agriculture. It recently signed 100-year leases with four urban communities (Phoenix, Scottsdale, Goodyear, and Peoria) and will use the revenue to support other forms of economic development on the reservation.

The tribe faces obstacles in realizing its dreams of re-establishing itself as the new breadbasket of the Southwest. The Bureau of Indian

Affairs (BIA) oversees approval of economic development projects on the reservation, including land use. In 1921, the BIA reorganized the most productive land from traditional communal to individualized ownership. This process resulted in 4,894 10-acre irrigable "A" allotments and 4,894 non-contiguous grazing "B" allotments, with each tribal member receiving two 10-acre allotments, one irrigable and one non-irrigable. The BIA now holds these allotments in trust for tribal members while the remaining 274,400 less productive acres are in communal ownership by the tribal council. The system of landownership and the BIA have made it difficult to lease land for large-scale commercial operations. Few tribal members have wills that designate how to divide their allotments upon their death. As a result, each surviving heir receives an equal share of land. Multiplied over five or six generations, there are now hundreds of landowners with joint interest in each 10-acre parcel. The agency responsible for oversight of the leases lists 121,136 shares owned by tribal members, nearly two-thirds of them with less than a 2% interest (DeJong 2014). The fragmentation of land ownership complicates efforts to organize lease agreements with producers. The BIA's method for managing leases further limits development as it allows only a five-year term limit, making it hard for growers to amortize the costs of improvements and infrastructure.

After a century of litigation and negotiation of water rights, the GRIC has not yet realized its dream to redevelop its agricultural past, and now confronts the reality that irrigation agriculture may not be the highest and best use of its precious water supplies moving forward (Bark and Jacobs 2009). Traditional communal land ownership rules have not been reconciled with legal norms of individual ownership, and future development is in limbo. Despite these challenges, GRIC holds the largest share of unused water rights in one of the fastest growing large urban regions in the US. The People of the River will influence the amount and direction of urban growth, the preservation of the region's agricultural heritage and cultural landscape, and potential for maintaining an agricultural buffer as a hedge against future drought conditions.

5.7 Conclusions

This chapter tackled the importance of values in water management in cities, on farms, and in remote and highly urbanized regions where Indigenous peoples have practiced water-based occupations for thousands of years. Water is a place-based resource with deep cultural meaning for individuals and communities. Solving complex water and climate change problems requires deep understanding of what water means to people, beyond its economic and life-sustaining value. Recall Mikhail Gorbachev's warning at the start of this chapter that the distant alarm comes closer as people everywhere and everyday need more water. New strategies are required to integrate the meaning, purpose, and value of water into governance, science practice, and decision-making frameworks.

References

Abbott, David R. 2000. *Ceramics and Community Organization Among the Hohokam*. Tucson, AZ: University of Arizona Press.

Abu, Razak, Reed, Maureen. G. and Timothy D. Jardine. 2017. "Understanding a Hydro-socio-ecological System Through Two-eyed Seeing: Bringing Together Traditional and Scientific Knowledge to Understand Long-term Change in the Saskatchewan River Delta, Canada." Unpublished manuscript.

Anderson, Terry L. and Donald R. Leal. 2001. *Free Market Environmentalism*. London: Palgrave Macmillan.

Andrews, Evan. 2015. "Environmental Justice and Dam Management: A Case Study in the Saskatchewan River Delta." Master's thesis. University of Saskatchewan. 2015. https://ecommons.usask.ca/handle/10388/ETD-2015-12-2337. Accessed June 19 2017.

Bakker, Karen 2007. "The 'Commons' Versus the Commodity: Alter-globalization, Anti-privatization and the Human Right to Water in the Global South." *Antipode* 39(3): 430–449.

Bandler, Hans. 1995. "Water Resources Exploitation in Australian Prehistory Environment." *The Environmentalist* 15(2): 97–107.

Bark, Rosalind H., and Katharine L. Jacobs. 2009. "Indian Water Rights Settlements and Water Management Innovations: The Role of the Arizona

Settlements Act." *Water Resources Research* 45(5): W05417. https://doi.org/10.1029/2008WR007130. http://onlinelibrary.wiley.com/doi/10.1029/2008WR007130/pdf. Accessed September 28 2017.

Battiste, Marie. 2000. *Reclaiming Indigenous Voice and Vision.* Vancouver: University of British Columbia Press.

City of Phoenix Water Services Department. 2014. "Water Production and Water Use, 1990–2013." obtained through public records request.

City of Tucson. 2017. "About Tucson Water (Part Two)." https://www.tucsonaz.gov/water/about-us-2. Accessed June 18 2017.

Coulthard, Sarah. 2008. "Adapting to Environmental Change in Artisanal Fisheries—Insights from a South Indian lagoon." *Global Environmental Change* 18(3): 479–489.

DeJong, David H. 2007. "The Sword of Damocles? The Gila River Indian Community Water Settlement Act of 2004 in Historical Perspective." *Wicazo Sa Review* 22(2): 57–92.

DeJong, David H. 2014. "Navigating the Maze: The Gila River Indian Community Water Settlement Act of 2004 and Administrative Challenges." *The American Indian Quarterly* 38(1): 60–81.

Draper, Stephen E. 2008. "Limits to Water Privatization." *Journal of Water Resources Planning and Management* 134: 493–503.

Ducks Unlimited. 2017. "Saskatchewan – The Land of Living Skies." http://www.ducks.ca/places/saskatchewan/. Accessed September 17 2017.

Dumanski, Stacy, John W. Pomeroy, and Cherie J. Westbrook. 2015. "Hydrological Regime Changes in a Canadian Prairie Basin." *Hydrological Processes* 29: 3893–3904.

Economist 2017. "Maori rights in New Zealand: Water, Water Everywhere." Accessed on June 19 2017 at: http://www.encyclopedia.com/science/news-wires-white-papers-and-books/planning-and-management-history-water-resources.

European Association of Public Water Operators. 2017. "Our Values." http://www.aquapublica.eu/?Our-Values-62. Accessed June 17 2017.

Finger, Matthias and Jeremy Allouche. 2002. *Water Privatisation: Transnational Corporations and the Re-regulation of the Water Industry.* London and New York: Spon Press.

Funk, Cary and Brian Kennedy. 2016. "The Politics of Climate." Pew Research Center. http://www.pewinternet.org/2016/10/04/the-politics-of-climate/. Accessed August 16 2017.

Getches, David H. 2005. "Indigenous Peoples' Rights to Water Under International Norms." *Journal of International Environmental Law & Policy* 16(2): 259–294.

Gober, Patricia. 2006. *Metropolitan Phoenix: Place Making and Community Building in the Desert*. Philadelphia, PA: University of Pennsylvania Press.

Gorbachev, Mikhail. 2000. "Out of Water: The Distant Alarm Comes Closer." *Civilization: Magazine of the Library of Congress* November, 2000: 92.

Hanak, Ellen and Jelena Stryjewski. 2012. "California's Water Market, By the Numbers: Update 2012." Public Policy Institute of California. http://www.ppic.org/publication/californias-water-market/. Assessed August 16 2017.

Hassanzadeh, Elmira, Amin Elshorbagy, Ali Nazemi, Timothy A. Jardine, Howard Wheater, Karl-Erik Lindenschmidt. 2017. "The Ecohydrological Vulnerability of a Large Inland Delta to Changing Regional Streamflow and Upstream Irrigation Expansion." *Ecohydrology* 10(4): https://doi.org/10.1002/eco.1824.

Hidalgo, Juan Pablo, Rutgert Boelens, and Jeroen Vos. 2017. "De-colonizing Water. Dispossession, Water Insecurity, and Indigenous Claims for Resources, Authority, and Territory." *Water History* 9: 67–85.

Hiscock, Peter. 2008. *Archaeology of Ancient Australia*. London, UK: Routledge.

International Conference on Water and the Environment. 1992. "Dublin Statement on Water and Sustainable Development." Dublin, January 26–31. http://www.wmo.int/pages/prog/hwrp/documents/english/icwedece.html. Accessed June 17 2017.

Jackson, Sue, Poh-Ling Tan, Carla Mooney, Suzanne Hoverman, Ian White. 2012. "Principles and Guidelines for Good Practice in Indigenous Engagement in Water Planning." *Journal of Hydrology* 474: 57–65.

Jackson, Sue. 2006. "Compartmentalising Culture: The Articulation and Consideration of Indigenous Values in Water Resource Management." *Australian Geographer*, 37(1): 19–32.

Johnstone, Nick and Libby Wood. 2001. *Private Firms and Public Water: Realising Social and Environmental Objectives in Developing Countries*. London: International Institute for Environment and Development.

Keen, Ian. 2004. "The Language of Property: Analyses of Yolngu Relations to Country." In *Ethnography & the Production of Anthropological Knowledge: Essays in Honour of Nicolas Peterson*, edited by Yasmine Musharbash and Marcus Barber, Chapter 7. Canberra, Australia: ANU Press. http://press-files.anu.edu.au/downloads/press/p111611/pdf/ch077.pdf. Accessed September 28 2017.

Marcin, Tim. 2017. "What Has Trump Said About Global Warming? Eight Quotes on Climate Change as He announces Paris Agreement Decision."

Newsweek, June 1 2017. http://www.newsweek.com/what-has-trump-said-about-global-warming-quotes-climate-change-paris-agreement-618898. Accessed September 26 2017.

Navajo, Isaac A. 2012. "Water Rights: A Transformative Perspective on Water Rights and Indigenous Peoples." Master's Thesis, May 2012. Tempe, AZ: Arizona State University.

Paul, Andrew J. 2013. "Environmental Flows and Recruitment of Walleye (*Sander vitreus)* in the Peace-Athabasca Delta." *Canadian Journal of Fisheries and Aquatic Sciences* 70: 307–315.

Pomeroy, J. W., K. Shook, X. Fang, S. Dumanski, C. Westbrook, and T. Brown. 2015. "Improving and Testing the Prairie Hydrological Model at Smith Creek Research Basin." Center for Hydrology Report No. 14. http://www.usask.ca/hydrology/papers/Pomeroy_et_al_2014.pdf. Accessed September 28 2017.

Safi, Michael. 2017. "Ganges and Yamuna Rivers Granted Same Legal Rights as Human Beings." *The Guardian*, March 21, 2017. https://www.theguardian.com/world/2017/mar/21/ganges-and-yamuna-rivers-granted-same-legal-rights-as-human-beings. Accessed July 25 2017.

Sarewitz, Daniel. 2004. "How Science Makes Environmental Controversies Worse." *Environmental Science & Policy* 7: 385–403.

Saskatchewan Farm Stewardship Association. 2017. "Economic Viability and Sustainability of Saskatchewan Agriculture." Accessed June 18 2017. http://www.saskfsa.ca/wpcontent/uploads/downloads/SaskFSA_Brochure.pdf.

Saskatchewan Water Security Agency. 2011. "25 Year Saskatchewan Water Security Plan." https://www.wsask.ca/Global/About%20WSA/25%20Year%20Water%20Security%20Plan/WSA_25YearReportweb.pdf. Accessed August 16 2017.

Saskatchewan Water Security Agency. 2013. "The 25-Year Saskatchewan Water Security Plan." http://iharf.ca/wp-content/uploads/2014/11/25-Year-Saskatchewan-Water-Security-Plan.pdf. Accessed June 9 2017.

Saskatchewan Water Security Agency. 2015. "Agricultural Water Management Strategy." https://www.wsask.ca/Water-Programs/Agricultural-Drainage-/Agricultural-Water-Management-Strategy/. Accessed June 9 2017.

Saskatchewan Water Security Agency. 2016. "Dry Lake Drainage Project." http://iharf.ca/wp-content/uploads/2014/11/25-Year-Saskatchewan-Water-Security-Plan.pdf. Accessed June 18 2017.

Saskatchewan Water Security Agency. 2017. "Agricultural Water Management Strategy." https://www.wsask.ca/Water-Programs/Agricultural-Drainage-/Agricultural-Water-Management-Strategy/. Accessed June 18 2017.

Smith, Norman D., Galina S. Morozova, and Martin R. Gibling. 2014. "Channel Enlargement by Avulsion-induced Sediment Starvation in the Saskatchewan River." *Geology* 42(4): 355–358.

Statistics Canada. 2016a. "Census of Agriculture, Total Area of Farms and Use of Farm Land, Canada and Provinces." http://www5.statcan.gc.ca/cansim/a26?lang=eng&id=40002. Accessed January 22 2018.

Statistics Canada. 2016b. "Census of Agriculture, Selected Crop Data, Canada and Provinces." http://www5.statcan.gc.ca/cansim/a26?lang=eng&retrLang=eng&id=0040003&&pattern=&stByVal=1&p1=1&p2=31&tabMode=dataTable&csid=. Accessed January 21, 2018.

Statistics Canada. 2017. "Census Profile, 2016 Census, Cumberland House Cree Nation 20." http://www12.statcan.gc.ca/census-recensement/2016/dp-pd/prof/details/page.cfm?Lang=E&Geo1=CSD&Code1=4718801&Geo2=PR&Code2=35&Data=Count&SearchText=Cumberland%20House%20Cree%20Nation%2020&SearchType=Begins&SearchPR=01&B1=All&GeoLevel=PR&GeoCode=4718801&TABID=1. Accessed January 22 2018.

Stokes, Bruce, Richard Wicke, and Jill Carle. 2015. "Global Concern about Climate Change, Broad Support for Limiting Emissions." Research Report. http://www.pewglobal.org/2015/11/05/global-concern-about-climate-change-broad-support-for-limiting-emissions/. Accessed August 16 2017.

Turner, Dale A. 2006. *This Is Not a Peace Pipe: Towards a Critical Indigenous Philosophy.* Toronto: University of Toronto Press.

United Nations. 2007. "United Nations Declaration of Rights of Indigenous Peoples." Resolution 61/295 of the General Assembly. http://www.un.org/esa/socdev/unpfii/documents/DRIPS_en.pdf. Accessed September 28 2017.

United Nations. 2009. "Water in a Changing World." UN Water Development Report 3. Paris: UNESCO and Earthscan http://unesdoc.unesco.org/images/0018/001819/181993e.pdf. Accessed August 16 2017.

US Supreme Court. 1908. "Winters v. United States, 207 U.S. 564 (1908)." 2017. https://supreme.justia.com/cases/federal/us/207/564/case.html. Accessed June 19 2017.

von der Porten, Suzanne and Rob C. de Loë. 2014a. "How Collaborative Approaches to Environmental Problem Solving View Indigenous Peoples: A Systematic Review." *Society and Natural Resources* 27(10): 1040–1056.

von der Porten, Suzanne and Rob C. de Loë. 2014b. "Water Policy Reform and Indigenous Governance." *Water Policy* 16: 222–243.

West Water Research. 2015. "West Water Research Announces 2014 Water Rights Price Index Results." *Water Market Insider* 03 2015. http://www.waterexchange.com/wp-content/uploads/2015/09/15-0916-Q3-WWInsider-fnl-LO-REV.pdf. Accessed August 16 2017.

6

Decision Making Under Uncertainty (DMUU)

6.1 Introduction

Uncertainties about climate and nexus issues stimulate new thinking about water planning. Building resilience involves the capacity to look ahead, ask "what if" questions about the future, remain flexible in the face of uncertainty, and seek out policies that provide good outcomes no matter what the future climate brings in terms of climate and water impacts. This chapter introduces the idea of wicked problems characterized by uncertainty, complexity, and controversy. Decision Making Under Uncertainty (DMUU) strategies are appropriate for managing wicked problems. The shift to DMUU calls for new modeling strategies, a new role for science in water planning, public engagement, knowledge mobilization, and learning to live with uncertainty.

6.2 Decision Making Under Uncertainty

DMUU acknowledges that many aspects of the future are unknowable and that predictions and forecasts represent only one of many possible futures. This is not a new idea, as futurists like Herman Kahn talked

P. Gober, *Building Resilience for Uncertain Water Futures*,
https://doi.org/10.1007/978-3-319-71234-5_6

about the limits of forecasting more than 50 years ago. In thinking about policy for nuclear deterrence, Kahn worked from multiple visions of the future (e.g., business as usual, total disarmament, an ever spiraling into catastrophe theme, and a happy ending to the arms race). He talked about a set of conditions that work together to bring about the happy-ending scenario. Nuclear war is a complex, uncertain, and highly contested problem. He noted that "there are no panaceas, no free lunches, no ultimate and complete solutions. But determined efforts by large numbers of responsible people to make a difference" (Kahn 1984, p. 53). Embedded in this view is the need to move toward public consensus and incorporate a range of viewpoints in futures thinking. Science is but one constituency in managing consensus for climate adaptation and resilience in the water sector.

Standard methods of risk analysis from a variety of disciplines, including game theory, economics, operations research, and statistical decision theory, have been successfully applied to policy problems for many decades. These methods are adequate for problems in which system behavior is predictable and probability distributions are accepted and widely used to assess risk. Increasingly, however, society faces problems such as climate change, sustainable development, the introduction of new technologies and treatments, and environmental management where uncertainties and complexity can lead to surprising and catastrophic outcomes.

Deep uncertainty characterizes situations in which analysts do not know or cannot agree upon the key drivers that will shape the future, probability functions that represent uncertainty, and how to value gains and losses from key outcomes (Lempert et al. 2003). The essence of non-stationarity in an environmental system is that the past is not an adequate guide to the future. DMUU embraces flexibility as a way to manage an uncertain future. Conservation scientists, for example, recognize the need to adjust management strategies in light of climate shifts, reassess as needed, and modify goals if appropriate (Stein et al. 2013). From a traditional perspective focused on actions to promote ecosystem persistence, the field is moving toward enabling the transition of ecosystems to new functional states (Millar et al. 2007). For example, a recent study of forests in the Greater Chicago Area found that 15% of the current trees have moderately high or high vulnerability to climate change. Knowledge about the viability of different species helps forest and park managers and homeowners to decide what trees to plant now for the climate in 2050 (Brandt et al. 2017).

DMUU problems are "wicked" in the sense that they have human and technical dimensions (Chap. 2), link to climatic uncertainties (Chap. 3), exhibit interdependencies (Chap. 4), and are often highly contested (Chap. 5). Wicked problems are hard to solve because traditional technical approaches often ignore the values, perspectives, and local knowledge that people bring to problem solving. Addressing wicked problems requires collaboration, deliberation, and debate about the problem definition and acceptable solutions. "Wicked problems are managed, debated and constantly renegotiated rather than solved" (Head 2010, p. 10). They require better knowledge of interdependences and instruments for change and policy outcomes in a complex system. Also required are social processes that build collaboration among stakeholders to achieve a common understanding of the problem or at least respect for alternative problem definitions.

DMUU involves the search for robust strategies instead of utility maximization (picking the one best solution from a finite set). Robust strategies perform reasonably well when compared to alternatives across a range of plausible futures and evaluated using the many value systems held by different parties to the decision (Lempert et al. 2003 p. xiv). A viable strategy in the face of uncertainty is to hedge, in other words, to enhance flexibility to respond later if the need arises. From a water company's perspective, this may involve securing water supplies for the future, banking groundwater in case of future drought, or making market-based agreements contingent upon future drought conditions. Shaping actions may involve changing diets to reduce water consumption or altering the character of the built environment (e.g., reducing property lot size, increasing housing density, or removing irrigated turf grass) to reduce urban residential water demand (Gober et al. 2016a; Shandas and Parandvash 2010; Chang et al. 2010; House-Peters et al. 2010). Signposts are observations that warn of the need to change course to avoid negative outcomes. Monitoring is an important feature of DMUU as it is critical to recognize how conditions are changing and signal the need to amend plans.

Decision-makers and scientists practice DMUU in a variety of ways, both qualitatively in conceptualizing alternative visions of the future and quantitatively with modeling approaches. DMUU is defined less by a rigid set of steps and more by a change in the policy question from what is most likely to happen in the future to what are the ramifications of actions that can be taken now, given that we cannot predict the future (Lempert and Schlesinger 2000). Water managers are often more interested in knowing the future consequences of policy decisions over which they have control than potential climate change impacts over which they have little perceived control.

6.2.1 Robust Water Policies in Phoenix

Future supply and demand can unfold in an unplanned way, or a city can take action to avert crisis and mitigate risk. Gober et al. (2010) examined the consequences of policy decisions over a range of climate change scenarios for Phoenix, asking about the robustness of different policy strategies. A supply-oriented strategy involves 40% recycling of indoor water use, and a demand strategy involves setting limits on groundwater withdrawals (Table 6.1). Rows in Table 6.1 represent different policies with respect to supply, either no recycling or 40% recycling starting immediately. The 40% recycling means that the city reuses 40% of indoor residential water for outdoor landscaping, golf courses, or power plant cooling. The columns represent four different approaches to demand management, specifically, imposing constraints on water use to achieve

Table 6.1 Summary of illustrative policy alternatives

	4 Demand management policies			
2 Supply policies	No safe yield	Year 2008	Year 2020	15 MAF (18.5 MML) drawdown
No recycling	S-none, R-none	S-2008, R-none	S-2020, R-none	S-15MAF (18.5 MML), R- none
40% recycling	S-none, R-40%	S-2008, R-40%	S-2020, R-40%	S-15MAF (18.5 MML), R-40%

Note: MAF million acre-feet, *MML* million megaliters

safe yield and groundwater sustainability. In this context, sustainability means that there is no net drawdown of groundwater, averaged over a five-year period. A "No Safe Yield" policy imposes no constraints on water usage. "Year 2008" imposes a policy in 2008 that requires groundwater withdrawal to be in balance with recharge over a five-year period. "Year 2020" imposes the same policy but waits until 2020. A "15 MAF (18.5 MML) Drawdown" is a contingent or adaptive policy that imposes the five-year safe yield water use policy only when the cumulative drawdown of groundwater reaches 15 MAF (18.5 MML). Thus, there are eight different policy alternatives shown that combine supply and demand management policies. These eight policies use a shorthand notation system where "S-2020, R-none" represents a five-year sustainable demand policy implemented in 2020, and there is no recycling imposed.

Results of 400 simulations show that the two policies that impose sustainable demand in 2008 (S-2008, R-none and S-2008, R-40%) limit drawdown and increase in water levels in the aquifer when there are significant climate-induced reductions in river flows (Fig. 6.1). Restrictions on demand are robust to climate change in the sense that implementing sustainable demand restrictions immunizes the city from very unfavorable future climate conditions. Recycling further limits

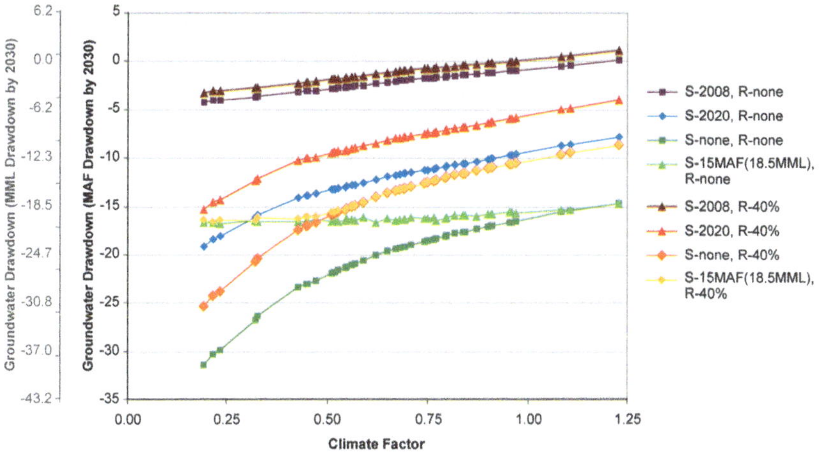

Fig. 6.1 Sensitivities of groundwater levels to water reuse and sustainable demand

consumption and maintains aquifer levels. Failure to impose demand constraints exposes the city to vulnerabilities from moderate to severe declines in water supplies due to climate change. Recycling mitigates groundwater drawdown only under small to moderate declines in river flows. Recycling alone does not prevent significant drawdown of the aquifer under severe climate conditions. The analysis demonstrates the value of precautionary action in the face of climatic uncertainty and points to greater importance of demand management over water recycling over the long term under the most negative climate change conditions (Gober et al. 2010).

6.3 Exploratory Modeling

Futures analysis often uses simulation models to represent how complex human–natural coupled systems respond to biophysical changes and policy decisions. Bankes (1993) distinguished between consolidative modeling that uses known facts to replicate an actual system and exploratory modeling in which models investigate the consequences of varying assumptions and hypotheses about the system and its future dynamics. The former uses optimization and prediction, while the latter acknowledges that not all relevant and important information is available. Exploratory modeling is appropriate for situations in which there is a high level of system complexity—where non-linear behaviors and feedbacks may result in unintended consequences and potentially catastrophic events. The search for an optimal solution will not reveal outcomes with catastrophic consequences nor will it necessarily reveal a path to avoid such consequences.

Exploratory models can be both scientific tools to investigate system behavior and communication devices to promote social learning about the system at hand. Ideally, they evolve iteratively to incorporate new information. Exploratory modeling does not aim to perfectly replicate the water system or predict in the face of a non-stationary future. Rather it aims to characterize essential features, identify critical vulnerabilities, and support policy discussion.

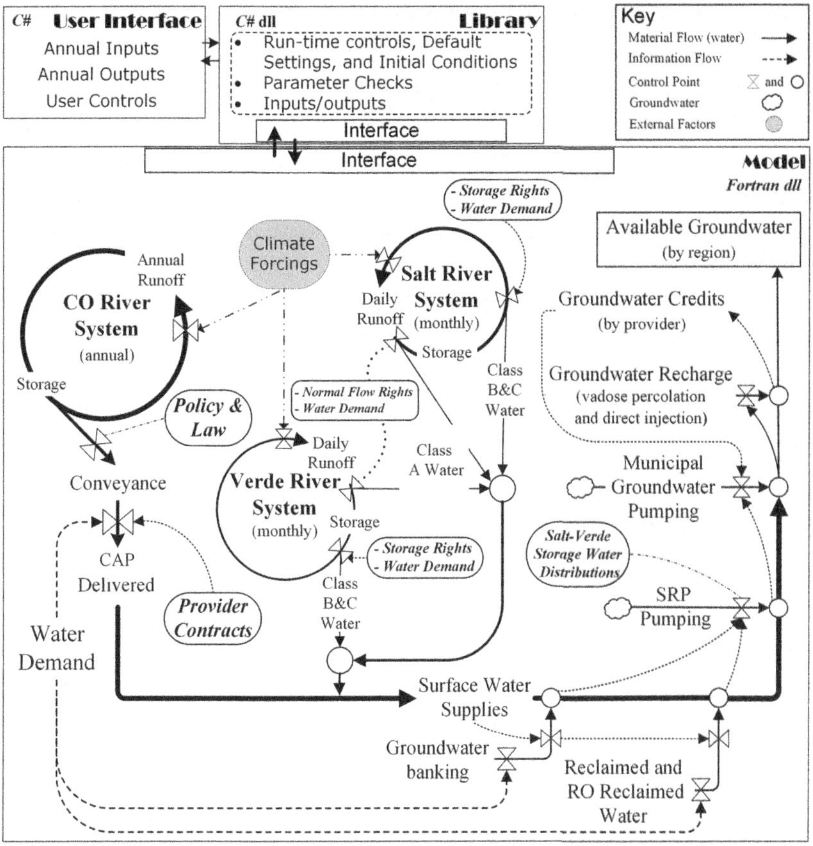

Fig. 6.2 WaterSim 5. David A. Sampson/Decision Center for a Desert City/Arizona State University

6.3.1 WaterSim 5

WaterSim 5 explores long-term water planning and management policies for surface and groundwater for metropolitan Phoenix (Fig. 6.2). It represents the modification of earlier versions developed over time by the Decision Center for a Desert City at Arizona State University (Gober et al. 2011; Sampson et al. 2011; Sampson et al. 2016). It consists of several sub-models that work together to characterize water supply and demand conditions in the Phoenix region. The current model focuses on urban water demand (indoor and outdoor) and uses provider-specific representations of water supply. It includes water treatment facilities and residential, commercial, and industrial water use from supply source to waste treatment and water reuse. The model uses estimates based on historical data to represent reservoir operations for three surface water sources (Salt, Verde, and Colorado Rivers) in Phoenix. It uses FITFIR water rights for local providers to allocate water from the Salt and Verde Basins immediately upstream from Phoenix. Groundwater estimates from the Arizona Department of Water Resources (ADWR) initialized available provider-level credits based on provider use in 1980, and then allowed providers to accumulate additional credits by recharging extra surface water into the aquifer. Providers use credits to meet the theoretical standard of a 100-year assured supply to enable land development, as discussed in Chap. 3.

6.3.2 Anticipatory Water Modeling for the Urban Southwest

Today's Phoenix builds on the remnants of an ancient civilization that lived along the banks of the Salt and Verde Rivers from 0 AD to 1250 AD. At their peak, the Hohokam numbered some 40,000 and developed a complex and productive society based on irrigation agriculture. They disappeared sometime in the thirteenth century. Archeologists surmise that they were unable to adapt to an unusually dry period during the Medieval Period of the twelfth century (Abbott 2000; Redman 1999). The name for modern Phoenix comes from the mythical Phoenix bird that rose from its own ashes to create a new civilization (Luckingham 1989). Thus, the ability to adapt to climate change impacts has deep roots in the cultural history of modern Phoenix.

Modern agricultural development during the late nineteenth century used remnants of the Hohokam irrigation canals. The region was the third most productive agricultural county in the nation in 1950 (Gober 2006). Post-World War II urbanization relied on upstream water from the Salt and Verde River Basins, augmented by construction of the Central Arizona Project (CAP), a 336-mile (541 km) aqueduct to bring Colorado River water from Parker, Arizona, to the Central Arizona cities of Phoenix and Tucson. CAP opened in 1985, providing the Phoenix area with an additional surface water source. The region also relied on groundwater from deep alluvial aquifers when surface water supplies were limited. One vision of a sustainable regional water system in this desert system uses groundwater as a bank to store water in times of plenty for use when surface supplies are low. In addition, artificial recharge of CAP water allows large volumes of Colorado River water to be stored to account for natural climatic variability.

Recent water supplies to metropolitan Phoenix have been robust to drought conditions. Sources are diverse; steady retirement of agricultural lands transfers water to urban uses; and users can bank excess water deliveries from the Colorado River via CAP for future use. A major concern for the future, however, is that climate change will reduce flows into Lake Mead, triggering reduced allocations of Colorado River water supplies. As part of the negotiations associated with CAP, Arizona agreed to guarantee California's 4.4 million acre-feet (5.4 billion m^3) a year as a priority over the CAP entitlement (Colorado River Users Association 2017). In other words, CAP absorbs any water shortage to Lower Basin States (California, Arizona, and Nevada) before supplies are curtailed to California and Arizona users along the Colorado River. Lake Mead water levels trigger officially declared shortage. Arizona's allocations are reduced by 320,000 acre-feet (394 million m^3) if Lake Mead water levels fall below 1075 feet (328) meters above sea level, 400,000 acre-feet (493.4 million m^3) at 1050 feet (320 meters), and 420,000 (518.1 m^3) at 1025 feet (312 meters) (US Department of the Interior, Bureau of Reclamation 2007). Near-term shortage would curtail CAP water supplies used to bank water to ensure 100-year assured supplies for new development to occur.

The eyes of Central Arizona water managers are thus fixed on Lake Mead water levels. Levels have hovered around 1075 feet, the first critical threshold, since 2014 (Fig. 6.3). The bathtub ring at the base of Mead is a powerful visual symbol of climate change impacts and falling lake levels with significant ramifications for regional water supplies (Fig. 6.4).

Uncertainties about climate change and drought conditions are set within an urban context of rapid urban growth and development. Between 2010 and 2015, the Phoenix urban region added about 100,000 new residents annually. Projections show it reaching eight million by 2050 (Arizona Department of Administration 2017). In recent years,

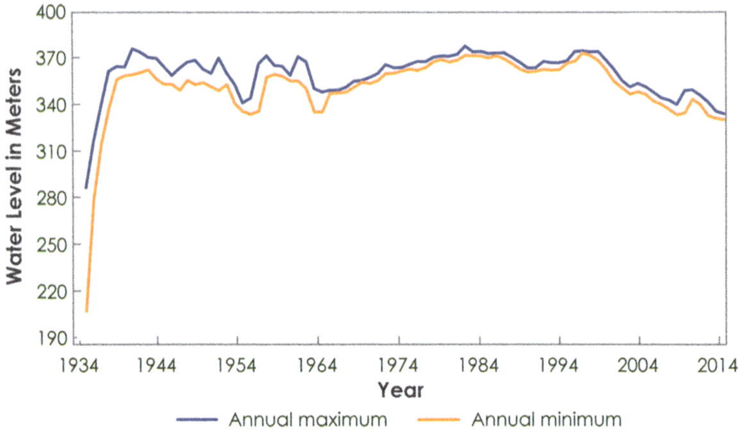

Fig. 6.3 Water levels in Lake Mead, 1934–2016. Bureau of Reclamation/https://www.usbr.gov/lc/region/g4000/hourly/mead-elv.html

Fig. 6.4 Bathtub rings in Lake Mead signal low water levels. Delta/OFF/Shutterstock

declines in per capita urban water use buffered the effects of an increasing population, but there are limits to the potential for water conservation to compensate for growth as indoor use has steadily fallen and new construction favors higher-density development with smaller gardens, fewer pools, and less outdoor use.

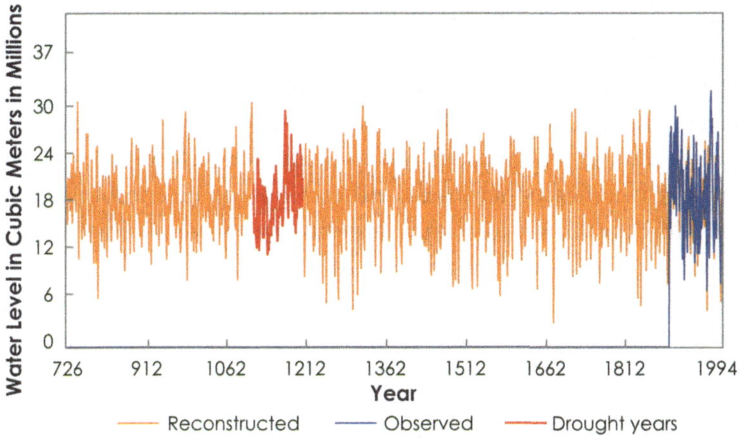

Fig. 6.5 Reconstructed flows of the Colorado River at Lees Ferry, 762-2005. Meko et al. 2017/http://treeflow.info/upco/coloradoleesmeko.html

A DMUU exercise explored the impact of varying policy decisions on groundwater sustainability in Phoenix under conditions of mega-drought (Gober et al. 2016b). Instead of using a downscaled climate model to predict the future, this study used prehistorical tree-ring data dating from 762 AD to 2005 AD to identify a particularly dry period in the record during the twelfth century commonly known as the Medieval Drought (Meko et al. 2007; Woodhouse et al. 2010) (Fig. 6.5). The research explored how Phoenix would fare if Medieval Drought conditions were to reoccur. There is no need for public discussion of climate models or their uncertainties. The Medieval Drought happened, the Hohokam moved away, and the modern city rests on its irrigation prehistory. In keeping with a DMUU approach, emphasis is on the effectiveness of policy instruments rather than on assessing climate change impacts. The specific research question is whether Phoenix would be able to survive another mega-drought and maintain the legal standard of a 100-year supply of groundwater. Secondary questions explore the effectiveness of various policy options as precautionary measures.

Qualitative analysis of community discussion yielded five potential policy tools for the quantitative analysis using WaterSim 5 (White et al. 2015). These were (1) population growth management, (2) municipal and industrial water conservation, (3) groundwater banking and recharge, (4) direct potable reuse of Reverse Osmosis (RO) reclaimed water, and (5) supply augmentation with a desalination plant. Arizona's Groundwater Management Act (GMA) of 1980 requires a 100-year supply of sustainable water resources to approve new residential development. This study determines whether policy tools, either alone or in combination with others, allow the region to achieve groundwater sustainability in the face of significant climate constraints.

Sensitivity analysis investigates potential changes in groundwater levels under the five policy scenarios separately (Fig. 6.6a–e). A no-growth scenario produces close to 100 years of available groundwater supply. Phoenix could, in other words, maintain a sustainable 100 years of available supply under mega-drought conditions, but would need to limit growth completely, assuming a 1% conservation rate and current demand and supply management through 2060 (Fig. 6.6a). Population growth has a linear effect on groundwater supplies—more people demand more water, and if it is not available from surface supplies (as would be the case under mega-drought conditions), the region draws groundwater. Similarly, conservation reduces water demand, resulting in a linear response in groundwater levels. At an assumed level of 75% of projected growth, even conservation rates of 1.5% per year would not be sufficient to support the 100-year standard of available supply (Fig. 6.6b). Years of supply would range from 80 to 85 years, depending upon the conservation rate. Reuse extends available water supplies to ~83 years at the end of the simulation assuming 75% growth and 1% conservation (Fig. 6.6d). One reason for the leveling out after 2030 is that wastewater effluent, the source of RO reclaimed water, is primarily dependent on indoor water use. As indoor use becomes more efficient, water available for reuse will fall. Water banking produces relatively little change in the overall trend in years of adequate groundwater, but banking moderates the inter-annual variability in supply (Fig. 6.6c). Augmentation (Fig. 6.6e) produces a non-linear response in years of groundwater supplies. It takes the variability out of groundwater supplies, but does not extend years of supply to 100 years. Adding 7% of total regional demand to the available water supply from an outside source results in a much larger response than moving from 7% to 21%.

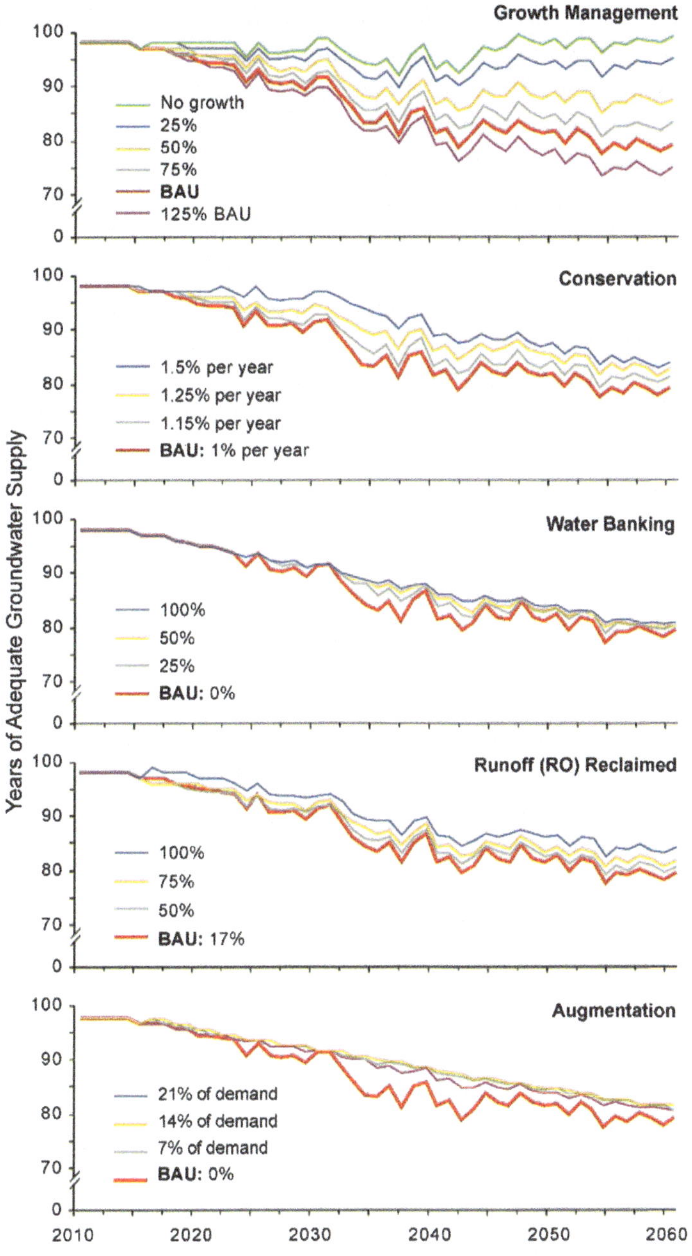

Fig. 6.6 Sensitivity of years of available water to drought-mitigation policies under mega-drought conditions

It is not likely that the region would implement the proposed policies separately. Rather, it would adapt a suite of adaptive strategies to mitigate shortages in anticipation of climate change impacts. Growth management, conservation, and reuse improve the system's capacity to meet the long-term needs of growth while banking and augmentation enhance reliability (Fig. 6.7). Assuming population growth at 75% of projected levels, there would be 82 years of available groundwater at the end of the simulation in 2060. This scenario exposes significant risk of dipping below 60 years of available supply, especially after 2030 (Fig. 6.7a). Adding conservation of 1% per year reduces uncertainty associated with the years of available supply and raises the mean years of available supply from 82 to 89 (Fig. 6.7b). Water banking adds little to years of available supply, but significantly reduces the risk of falling below 90 years (Fig. 6.7c). It is a mechanism for modulating the effects of inter-annual variability, not for extending years of supply. Reuse does not reduce the risk of falling below 90 years, but it does extend end-of-simulation supply to 94 years (Fig. 6.7d). Augmentation adds another three years to available supply and reduces risk of potential shortages in years of low surface flows.

It is possible to support growth rates of 75% of projections in Phoenix and maintain a sustainable groundwater supply under mega-drought conditions through 2060, but the region would require the full arsenal of infrastructure and conservation policies. Implementation would need to occur now to guarantee effectiveness in the mid-term. The Phoenix region does not currently have the physical infrastructure in place to recover and bank all excess surface water. Furthermore, reclaimed water and coastal desalination facilities require a decade or more to plan and finance. Conservation theoretically can be implemented more quickly, but there are cultural and lifestyle issues to be addressed as Phoenix transforms culturally from oasis to desert landscapes (Larson et al. 2011a; Neel et al. 2014; Gober et al. 2016a).

The policies chosen for this analysis and the order in which they were included in the model are arbitrary. The analysis demonstrates DMUU thinking and strategy. The goal is to ask "what if" questions, not to plot the most efficient or desirable adaptation strategy. The region faces uncertainties about future water supplies stemming from climate change

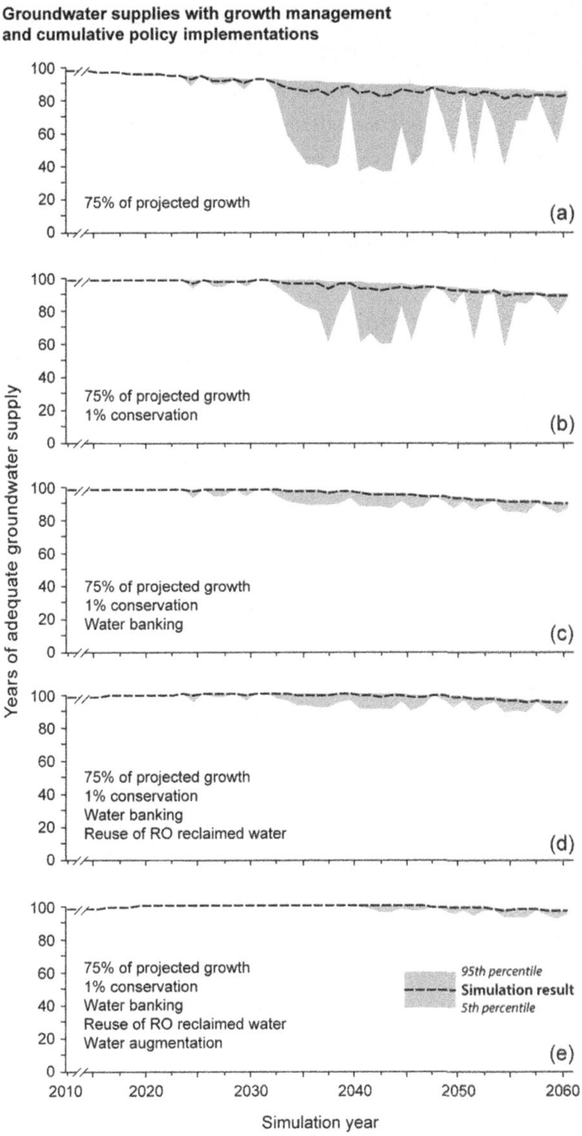

Fig. 6.7 Groundwater supplies with growth management and cumulative policy implementations. **(a)** growth management GM, **(b)** GM + conservation C, **(c)** GM, C + water banking WB, **(d)** GM, C, WB, + reuse RU, **(e)** GM, C, WB, RU + desalination

impacts and the vulnerability of its Colorado River water supplies due to the CAP agreement with California. Political leaders can wait and see what happens to Lake Mead water levels or take precautionary action now in a variety of forms to ensure adequate water supplies to a larger population living in a changing climate.

6.3.3 Scenario Planning in Las Vegas

Scenarios are another mechanism for exploring the future. They are consistent stories about the future for systems that are too complex to predict. They are constructed for different time periods (10 years to 1000 years), spatial units (small villages to global climate models), and purposes (as inputs to modeling experiments or as capacity building exercises for decision support) (Wiek et al. 2013). They are a way to generate anticipatory knowledge to help prepare for an uncertain future. They cover the range of plausible futures, including rare but potentially devastating events. It makes sense to anticipate the widest possible range of future conditions and decide whether and how to plan for them, given the limitations of current scientific predictions. Scenarios often include a worst-case option to focus public attention on the need to act, despite wide bands of uncertainty that surround many water planning situations.

Las Vegas produced scenarios as a means of future water planning. The city is dependent on the Colorado River for 90% of its current water supplies. It was a mere speck on the map when Colorado River allocations were set in 1922. The Colorado River Compact apportioned 0.3 million acre-feet (370.0 million m^3) of water to Nevada (compared to 4.4 million to California and 2.8 million to Arizona). At the time, Nevada had no significant agricultural or industrial users, and groundwater seemed plentiful (Southern Nevada Water Authority 2015). The situation changed dramatically, however, and by the latter part of the twentieth century, Las Vegas was a large and rapidly growing metropolitan area specializing in gaming and tourism. The urban region doubled in size between 1990 and 2000 from 741,000 residents to 1.4 million (US Bureau of the Census 2006). High outdoor use and water-intensive lifestyles, along with a large visitor population, increased pressure on the city's Colorado River allocation. By the end of the century, there were concerns about its capacity to support further growth. Las Vegas implemented its first water conservation program (Operation Desert Lawn) in the late 1990s including ordinances to restrict landscape irrigation during the hottest hours of the day.

Efforts to reduce per household demand intensified more recently as the metropolitan population grew to 2.0 million in 2010 and 2.2 million in 2016 (US Census 2016). Conservation programs included incentives to remove turf grass and replace it with water-efficient landscaping, rebate programs for pool covers, smart irrigation controllers and rain sensors, financial incentives to owners of multi-family housing and commercial establishments to install water-efficient fixtures and appliances, and regulations to limit outdoor water use for watering lawns and washing cars. Regulations forbid turf grass in the front yards of new homes and limit them to 50% of the backyard. They also ban commercial misters and the use of ornamental fountains and other water features in most jurisdictions. Conservation efforts include education and public outreach programs. Total gallons per capita per day (GPCD) in Las Vegas declined from 348 (1.32 m^3) in 1989 to 205 (.78 m^3) in 2014 (Southern Nevada Water Authority 2015), an impressive 43% decline. Water use in Las Vegas remains high, however, compared to 117 (.443 m^3) in Tucson (City of Tucson 2017), 163 (.62m^3) in Phoenix (2014) (City of Phoenix Water Services Department 2014), 113 (.428 m^3) in Los Angeles (2015), and 83 (.324 m^3) in San Francisco (2014) (The San Francisco Indicator Project 2014).

Las Vegas also faces uncertainty with respect to estimated water demands. The 2007–2008 global recession hit the city hard (Fig. 6.8). Tourism and gaming revenues fell, and unemployment rates spiked. New residential and commercial development projects came to a halt, and home foreclosures flooded the real estate market. Although the region rebounded, there is growing disquiet about predictions of future growth and water demand, given the city's economic structure and recent history (Southern Nevada Water Authority 2015).

The city's most recent long-term water resource plan in 2015, "Water Resources Plan 2015," relies on scenarios to address demand and supply uncertainties. It uses a mid-term planning horizon (2065) and grapples with the costs and trade-offs associated with continued growth. Participants in the planning process were particularly concerned about

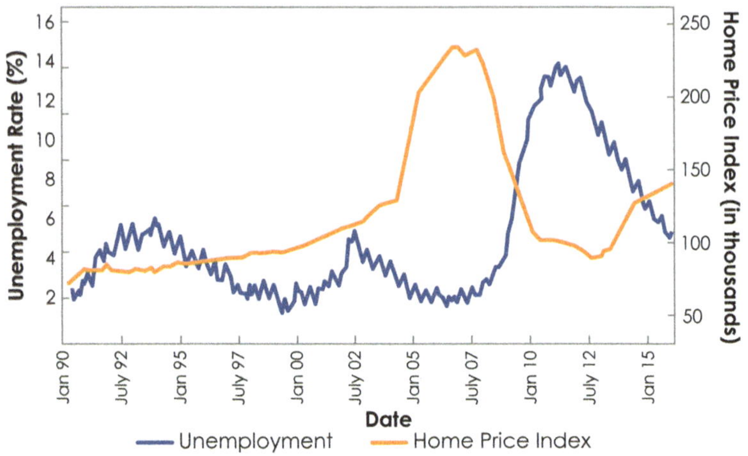

Fig. 6.8 Unemployment rates and home price index in Las Vegas, 1990–2017. Federal Reserve Economic Data/https://fred.stlouisfed.org

the city's ability to access water from Lake Mead if water levels fall below 1000 feet (305 m)—once an unthinkable idea, it is now a subject of serious public discussion. The city had intake stations at 1000 and 1500 feet (305 and 457m) above sea level. Lake levels hovered between 1072 and 1088 feet (327 and 333 m^3) for most of 2016 and 2017, and climate research suggests the dire possibility of levels falling further (Cook et al. 2015). The Planning Committee characterized the risk of Mead's elevation falling below 1000 feet (305 m) as "unacceptable" and recommended building a new intake facility at 860 feet (262 m). The price of the infrastructure now was worth the cost of reducing the risk of Lake Mead falling below 1000 feet sometime in the future. This precautionary action occurred with public support and articulation of water levels in Mead and growing recognition that predicting future water demand is a problem of deep uncertainty.

The plan also included a diverse set of water resource options to meet future needs, some of them to be deployed now and others later if needed. They include groundwater banking, additional conservation, building a desalination plant in Mexico, and so-called intentionally created surpluses (Las Vegas has rights to water in Nevada near the Colorado River that it may develop as needed, allowing it to flow into Lake Mead).

The city explored "what if" policy choices under different supply and demand conditions. Modeling results revealed the consequences of various policy choices. Even in the best-case scenario, normal supplies from the Colorado River combined with moderate growth in demand would require new water sources by 2065. The "increasing shortage" future scenario requires the city to use temporary resources (e.g., banked water, return-flow credits, intentionally created surpluses) until 2040 and then look to a portfolio of expensive and socially contested water resources (desalination, litigation, water transfers) to meet increased demand in 2065.

Skeptics argue that these exercises delay inevitable decisions about additional conservation and growth restrictions and provide a false sense of security to prolong economic development. Prices are too low, they say, to stimulate needed levels of conservation, and there is inadequate oversight of over-irrigators. The various scenarios and adaptive strategies downplay the fact that Las Vegas relies almost exclusively on a single vulnerable water source (Glionna 2014). The city has, however, thought seriously about future climate change impacts and anticipatory actions to deal with them.

6.3.4 Futures Planning in Australia's Murray-Darling Basin

The semi-arid Murray-Darling Basin (MDB) in Australia also confronts the challenge of non-stationarity with a fully allocated water system. It covers 14% of Australia's land area, contains around 10% of the country's population (~2 million), and produces 39% of the value of Australia's agricultural production (Murray-Darling Basin Authority (MDBA) 2010). The region includes 65% of Australia's irrigated agricultural lands and large urban regions, counting Brisbane and the Sun and Gold Coasts (Fig. 6.9). Indigenous hunters and gatherers occupied the lands for millennia before British colonists developed commercial agriculture in the late nineteenth century for internal consumption and export (Weir 2011). Large engineering projects began in the 1920s as settlers constructed dams, weirs, locks, canals, and pipes to divert water to irrigation

Fig. 6.9 Agricultural landscape along the Murray River near Murray Bridge

agriculture and for use by rural towns and communities. Water diversions led to reduced flows to floodplains and wetlands and altered the character of downstream ecosystems and hydrological processes, and changed the timing of peak flows from autumn and winter to spring and summer.

MDB's flat topography and limited groundwater outflows make it prone to salinization (Abel et al. 2016). The closed nature of the Murray Basin results in strong interactions between surface water and groundwater. The sediment layer in which groundwater can be stored is thin, and thus provides relatively small storage capacity. When recharged from surface water irrigation since the time of European settlement, the recharge could not escape and water tables rose. Mountains of salt that had been dormant for 18,000 years were reactivated, moving salt into water and onto the land. The fact that groundwater drains internally rather than toward the sea limits the capacity to flush saline groundwater out of the system. Salinity can fall only slowly as groundwater levels decline (Williams 2011).

Twentieth-century development of the MDB assumed that irrigation offered effective "drought proofing" for dryland agriculture (Quiggin 2011). By the 1980s, there was increasing concern about the impacts of surface water extractions on the health of the river and its ecosystems. Assessments in the 1990s revealed that diversions already exceeded sustainable levels by one-third (Williams 2011). Water quality problems appeared in the form of excess salinity (Quiggin 2011). Efforts to elicit cooperation and coordination across the four states (South Australia, Victoria, New South Wales, and Queensland) included in the basin began with the creation of water markets in 1994 with the Council of Australian Governments Communique that led to the National Water Initiative (NWI) in 2004. The so-called Cap placed restrictions on future annual surface extractions at 11,600 GL and implemented water trading. However, NWI failed to regulate groundwater or address the fundamental problems of over-allocation, inadequate environmental flows, and degraded ecosystems (Williams 2011; Quiggin 2011).

The basin thus entered the Millennium Drought (2000–2009) unprepared to face significant water shortages and challenge basic assumptions about the efficacy of irrigation agriculture, drought policy, risk management, and the health of the river. Drought put many farmers into financial stress, and rural communities deteriorated. The assumption that irrigation agriculture would be able to weather severe droughts better than dryland agriculture proved not to be the case. The health of the river and its aquatic ecosystems suffered. In response, the federal government passed the Water Act of 2007 that implemented (in 2012) a comprehensive water reform process. The Act set a sustainable diversion limit of the water of the Murray-Darling Basin for human use. Major crops include rice, cotton, dairy, and horticulture (Murray-Darling Basin Authority 2010). Assessments determined the sustainable limit, and it was set at 10,873 gigaliters (GL) in 2012. This limit required MDBA to find a way to return 2750 GL of surface water to the river system to ensure healthy ecosystems and a sustainable future for generations to come. This figure was at the low end of the recommended water return by scientific analysis of what it would take to achieve the environmental objectives of the Water Act.

Announcement of the plan on October 8, 2010, caused a maelstrom of bad publicity within the basin and nationally (Wahlquist 2011). The process set forth in the Act called for a top-down process without public consultation despite the fact that return flow would have significant effects on farmers, rural communities, and the environment. Plan development took three years with little communication coming from the MDBA. Farmers and their representatives took control of the narrative on announcement day, and framed the story as soaring food prices, significant losses to irrigated agriculture, demise of tens of thousands of jobs, and severe social and economic impacts on rural communities. Lost in the story were the initial problems of an over-allocated basin, difficult trade-offs between agricultural production and environmental flows, and 150 years of mismanagement. Mike Taylor, Authority Chairman, tried to argue that the Basin Plan used scientific analysis to determine the water requirement necessary to maintain and restore ecosystems and environment assets. The authority, in fact, had used the lower limit to minimize social and economic impacts. Mr. Taylor resigned on December 7, 2010, reflecting the difficult political problem of restoring health to the Murray-Darling River Basin; the need for public consultation before, during, and after the science and assessment processes; and the challenges of operationalizing and communicating the competing goals of a healthy river, food production, and viable communities.

As the Basin Plan unfolded between 2012 and 2017, it employed many core features of DMUU. First, it acknowledged that plans must be flexible and adaptive to account for the uncertainties of climate change and water resource management. Second, the approach recognized that the traditional methods of using historical data to anticipate the future were inadequate and "in a changing climate this approach becomes less and less reliable" (Murray-Darling Basin Authority 2017). Third, monitoring and evaluation covered a wide range of topics, including how people in irrigated agriculture adjust to reduced water for consumptive uses and how different levels of recovery affect river health. Fourth, the overall Basin Plan now includes input from people most affected by it at the local level in addition to cross-Basin goals and allocation targets. Fifth, scenario analyses of Basin communities in response to different climate conditions and policy interventions asked "what if" questions and revealed significant vulnerabilities, particularly in a cluster of Southern Basin communities.

6.4 Conclusions

DMUU ideas have been included explicitly and implicitly in water sector planning and design with the onset of climate change and other uncertainties (e.g., population growth, policy implementation, and public response to policy). Water problems described in this chapter (climate change, over-allocation, unsustainable use, unsustainable population growth, and poor governance) are wicked and thus in need of continuous negotiation and monitoring. They require exploratory modeling, anticipatory planning, hedging and shaping strategies to accommodate uncertainty, and public input. There is no one-size-fits-all set of procedures for DMUU, but rather a change in mindset from reducing uncertainty to adapting to it.

References

Abbott, David R. 2000. *Ceramics and Community Organization among the Hohokam*. Tucson, AZ: University of Arizona Press.

Abel, Nick, Russell Wise, Matthew Colloff, Brian Walker, James Bulter, Paul Ryan, Chris Normal et al. "Building Resilient Pathways to Transformation When No One Is in Charge: Insights from Australia's Murray-Darling Basin," *Ecology and Society* 21(2) (2016): 23. https://doi.org/10.5751/ES-08422-210223. Accessed June 22 2017.

Arizona Office of Economic Opportunity. 2017. "Population Projections." https://population.az.gov/population-projections. Accessed September 28 2017.

Bankes, Steve. 1993. "Exploratory Modeling for Policy Analysis." *Operations Research* 41(3): 435–449.

Brandt, Leslie A., David J. Nowak, Allison R. Bodine, Andrew Bell, Shannon Still, Patricia R. Butler, Andrea Dierich, Stephen D. Handler, Maria K. Janowiak, Stephan N. Matthews, Jason W. Miesbauer, Matthew Peters, Anantha Prasad, P. Danielle Shannon, Douglas Stotz, and Christopher W. Swanston. 2017 "Chicago Wilderness Region Urban Forest Vulnerability Assessment and Synthesis: A Report from the Urban Forestry Climate Change Response Framework Chicago Wilderness Pilot Project." U.S. Department of Agriculture, Forest Service, Northern Research Station

General Technical Report NRS-168. https://www.fs.usda.gov/treesearch-beta/pubs/54128. Accessed June 22 2017.

Chang, Heejun, G. Hossein Parandvash, and Vivek Shandas. 2010. "Spatial Variations of Single Family Residential Water Use in Portland, Oregon." *Urban Geography* 31: 953–972.

City of Phoenix Water Service Department. 2014. "Water Production and Water Use: 1990–2013." Obtained through public record request.

City of Tucson. 2017. "About Tucson Water (Part Two)." https://www.tucsonaz.gov/water/about-us-2. Accessed June 18 2017.

Colorado River Water Users Association. 2017. "Law of the River: The Colorado River Compact." https://www.crwua.org/colorado-river/uses/law-of-the-river. Accessed September 23 2017.

Cook, Benjamin I., Toby R. Ault, and Jason E. Smerdon. 2015. Unprecedented 21st Century Drought Risk in the American Southwest and Central Plains. *Science Advances*, 1(1): https://doi.org/10.1126/sciadv.1400082. http://advances.sciencemag.org/content/1/1/e1400082. Accessed June 23 2017.

Glionna, John M. 2014. "Drought—And Neighbors—Press Las Vegas to Conserve Water." *Los Angeles Times*, April 20, 2014. http://www.latimes.com/nation/la-na-las-vegas-drought-20140421-story.html. Accessed August 16 2017.

Gober, Patricia. 2006. *Metropolitan Phoenix: Place Making and Community Building in the Desert*. Philadelphia, PA: University of Pennsylvania Press.

Gober Patricia, Craig Kirkwood, Robert C. Balling Jr., Andrew W. Ellis, and Stephanie S. Detrick. 2010. "Water Planning under Climatic Uncertainty in Phoenix: Why We Need a New Paradigm." *Annals of the Association of American Geographers* 100(2): 357–372.

Gober, Patricia, Elizabeth A. Wentz, Tim Lant, Michael K. Tschudi, and Craig Kirkwood. 2011. "WaterSim: A Simulation Model for Urban Water Planning in Phoenix, Arizona, USA." *Environment and Planning B* 38(2): 197–215.

Gober, Patricia, Ray Quay, and Kelli L. Larson. 2016a. "Outdoor Water Use as an Adaptation Problem: Insights from North American Cities." *Water Resources Management* 30(3): 899–912.

Gober, Patricia, David A. Sampson, Ray Quay, Dave D. White, Winston T. L. Chow. 2016b. "Urban Adaptation to Mega-drought: Anticipatory Water Modeling, Policy, and Planning for the Urban Southwest." *Sustainable Cities and Society* 27 (2016): 497–504.

Head, Brian. 2010. "Wicked Problems in Water Governance: Paradigm Changes to Promote Water Sustainability and Address Planning Uncertainty." Urban

Water Security Technical Alliance, Technical Report No. 38. http://www.urbanwateralliance.org.au/publications/UWSRA-tr38.pdf. Accessed May 29 2017.

House-Peters, Lily, Benjamin Pratt, and Heejun Chang. 2010. "Effects of Urban Spatial Structure, Sociodemographics, and Climate on Residential Water Consumption in Hillsboro, Oregon." *Journal of the American Water Resources Association* 46(3): 461–472.

Kahn, Herman. 1984. *Thinking the Unthinkable in the 1980s*. New York: Simon & Schuster.

Larson, Kelli L., Amber Wutich, Dave White, Tischa A. Munoz-Erickson, and Sharon L. 2011a. "Multifaceted Perspectives on Water Risks and Policies: A Cultural Domains Approach in a Southwestern City." *Human Ecology Review* 18(1), 75–87.

Larson, Kelli L., Dorothy C. Ibes, and Dave D. White. 2011b. Water Risks and Policy Strategies: A Tripartite Conceptual Approach? *Environment and Behavior* 43(3): 415–438.

Lempert, Robert J., and Michael E. Schlesinger. 2000. "Robust Strategies for Abating Climate Change." *Climatic Change*, 45(3–4): 387–401.

Lempert, Robert J., Steven W. Popper, and Steven C. Bankes. 2003. *Shaping the Next One Hundred Years: New Methods of Quantitative, Long-term Policy Analysis*. Santa Monica, CA: Rand.

Luckingham, Bradford. 1989. *Phoenix: The History of a Southwestern Metropolis*. Tucson, AZ: University of Arizona Press.

Meko, David M., Connie A. Woodhouse, Christopher A. Baisan, Troy Knight, Jeffrey J. Lukas, Malcolm K. Hughes, and Matthew W. Salzer. 2007. "Medieval Drought in the Upper Colorado River Basin." *Geophysical Research Letters* 34 http://onlinelibrary.wiley.com/doi/10.1029/2007GL029988/epdf. Accessed September 23 2017.

Millar, Constance I., Nathan L. Stephenson, and Scott L. Stephens. 2007. "Climate Change and Forests of the Future: Managing in the Face of Uncertainty." *Ecological Applications* 17: 2145–2151.

Murray-Darling Basin Authority. 2010. "Guide to the Proposed Plan, Volume 2: Technical Background." https://www.mdba.gov.au/sites/default/files/archived/guide_pbp/Guide-to-proposed-BP-vol2-0-12.pdf. Accessed June 23 2017.

Murray-Darling Basin Authority. 2017. "2017 Basin Plan Evaluation." https://www.mdba.gov.au/basin-plan-roll-out/2017-basin-plan-evaluation. Accessed March 2 2018.

Neel, Rebecca, Edward Sadalla, Anna Berlin, Susan Ledlow, and Samantha Neufeld. 2014. "The Social Symbolism of Water-conserving Landscaping." *Journal of Environmental Psychology* 40: 49–56.

Quiggin, John. 2011. "Managing Risk in the Murray-Darling Basin," In *Basin Futures "Water Reform in the Murray-Darling Basin,"* edited by D. Connell and R. Q. Grafton, 313–326. Canberra: Australia National University E-Press.

Redman, Charles L. 1999. *Human Impacts on Ancient Environment.* Tucson: University of Arizona Press.

Sampson, D. A., Escobar, V., Tschudi, M. K., Lant, T., and Gober, P. 2011. "A Provider-based Water Planning and Management Model WaterSim 4.0 for the Phoenix Metropolitan Area." *Journal of Environmental Management*, 92: 2596–2610.

Sampson, D. A., Quay, R., and White, D. D. 2016. "Anticipatory Modeling for Sustainability in Phoenix." *Environmental Science & Policy* 55: 36–46.

San Francisco Indicator Project. 2014. "Water Use." http://www.sfindicatorproject.org/indicators/view/3. Accessed June 23 2017.

Shandas, Vivek, and G. Hossein Parandvash. 2010. "Integrating Urban Form and Demographics in Water Demand Management: An Empirical Case Study of Portland Oregon (US)." *Environment and Planning B: Planning and Design* 37: 112–128.

Southern Nevada Water Authority. 2015. "Water Resource Plan 2015." https://www.snwa.com/assets/pdf/wr_plan.pdf. Accessed June 23 2017.

Stein, Bruce A., Amanda Staudt, Molly S. Cross, Natalie S. Dubois, Carolyn Enquist, Roger Griffis, Lara J. Hansen, Jessica J. Hellmann, Joshua J. Lawler, Erik J. Nelson, and Amber Pairis. 2013. "Preparing for and Managing Change: Climate Adaptation for Biodiversity and Ecosystems." *Frontiers of Ecological Environment* 11((9): 502–510.

US Bureau of the Census. 2006. *Statistical Abstract of the United States: 2006 (125th Edition).* "Large Metropolitan Statistical Areas, 1990–2005." https://www.census.gov/library/publications/2005/compendia/statab/125ed.html. Accessed June 22 2017.

US Bureau of the Census. 2016. "Annual Estimates of the Resident Population: April 1 2010 to April 1, 2016." iiihttps://factfinder.census.gov/faces/tableservices/jsf/pages/productview.xhtml?pid=PEP_2016_PEPANNRES&src=pt. Accessed September 23 2017.

US Department of Interior, Bureau of Reclamation. 2007. "Colorado River Interim Guidelines for Lower Basin Shortages and the Coordinated

Operations for Lake Powell and Lake Mead." https://www.usbr.gov/lc/region/programs/strategies/RecordofDecision.pdf. Accessed September 23 2017.

Wahlquist, Åsa. 2011. "The Media and the Guide to the Basin Plan." *Water Reform in the Murray-Darling Basin*, edited by D. Connell and R. Q. Grafton, 115–134. Canberra, Australia: Australia National University E Press.

Weir, Jessica K. 2011. "Water Planning and Dispossession." In *Basin Futures "Water Reform in the Murray-Darling Basin,"* edited by D. Connell and R. Q. Grafton, 179–192. Canberra, Australia: Australia National University E Press.

White, Dave D., Lauren Withycombe Keeler, Arnim Wiek, and Kelli L. Larson. 2015. "Envisioning the Future of Water Governance: A survey of Central Arizona Water Decision Makers." *Environmental Practice* 17(1): (2015): 25–35.

Wiek, Arnim, Lauren W. Keeler, Vanessa Schweizer, and Daniel J. Lang. 2013. "Plausibility Indications in Future Scenarios." *International Journal of Foresight and Innovation Policy* 9(2/3/4): 133–147.

Williams, John. 2011. "Understanding the Basin and its Dynamics," In *Basin Futures "Water Reform in the Murray-Darling Basin,"* edited by D. Connell and R. Q. Grafton, 1-38 Canberra, Australia: Australia National University E Press.

Woodhouse, Connie A., David M. Meko, Glen M. MacDonald, Dave W. Stahle, and Edward R. Cook. 2010. "A 1,200-year Perspective of 21st Century Drought in Southwestern North America." *Proceedings of the National Academy of Sciences of the United States of America* 170(50): 21283–21288.

7

Urban Climate Adaptation

7.1 Introduction

Many cities around the world are now engaged in efforts to adapt to climate change. From New York and Rotterdam to Singapore and Durban, cities have embraced, to varying degrees, the relevance of uncertainty to long-term planning processes and the need for precautionary action. Water-related concerns include sea level rise, storm surges, effects of flooding on infrastructure, and the availability of water resources. Cities have emerged as the global focus for adaptation efforts because they have been able to capitalize on strategic leadership (e.g., Boris Johnson in London, Michael Bloomberg in New York City, and Richard Daley in Chicago). In addition, local-scale efforts tackle problems that local residents deem important and relevant to them personally, and cities provide a layer of governance that coordinates planning and decision-making (Hunt and Watkiss 2011).

P. Gober, *Building Resilience for Uncertain Water Futures*,
https://doi.org/10.1007/978-3-319-71234-5_7

7.2 Urban Adaptation

Cities are especially vulnerable to climate change impacts because they contain large concentrations of people. Some 54% of the world's population, four billion people, reside in urban areas (Population Reference Bureau 2017). Sixty percent of the world's 39 cities with a population of over five million are located within 100 km of coastlines, including 12 of the world's 16 cities with populations larger than ten million (Parry et al. 2007). Cities are highly dependent upon a complex network of infrastructure systems to provide energy, water, Internet service, sewage disposal, storm water drainage, health services, and transportation. These systems are often climate sensitive, and interconnections among them heighten their sensitivities to climatic variations. Adaptation efforts often stress a coordinated, cross-sector response to deal with interdependencies (Kirshen et al. 2008). Reduction of climate-related vulnerabilities in one sector usually lessens vulnerabilities in related sectors. For example, actions to improve water quality can help water supply, public health, and the environment. Retreat from flood-prone areas benefits the transportation system because fewer trips begin and end in flooded areas.

Urban land surfaces and biophysical conditions differ from the surrounding rural countryside, making them more vulnerable to climate change. Urban heat island (UHI) effects vary in time and space as a function of local meteorological conditions, location, and urban characteristics (Oke 1982). There is also an increase in the rate and volume of surface runoff from rainwater in cities (Mansell 2003). Urbanized surfaces replace natural ones that had provided shading, evaporative cooling, and rainwater interception and storage. Urban green spaces provide areas within cities where these functions can take place, and so-called green infrastructure is often a feature of urban adaptation strategies. Green space in urban areas provides cooler microclimates and reduces surface water runoff. A modeling study in Manchester, UK, showed that adding 10% green space in high-density residential areas and two centers would keep maximum surface temperatures at or below the 1961–1990 baseline until the 2080s (Gill et al. 2007). Adding vegetated land cover by 10% accelerates nighttime cooling in Phoenix by ~0.2°C/hour and reduces the rate of nighttime heating in Portland by 0.01°C/hour (Gober et al. 2012). That both cities are more sensitive to land cover changes than to climate conditions suggests that they can adapt to a warmer climate with urban redesign strategies.

Global urban adaptation plans exhibit interest and expertise in DMUU, although often without using the language of DMUU. Maunsell et al. (2008), for example, studied the effects of climate change on urban water supply infrastructure in Australia using various scenarios of the future. Potential impacts for major cities included a reduction in available water supplies, declining quality leading to higher treatment costs, and accelerated degradation and potential failure of the water distribution infrastructure. Goals do not always focus on the water sector, as there is broad acceptance of the interconnectedness of water, land, energy, and food. Reducing vulnerabilities in one sector often produces co-benefits in another. For example, green roofs lower both energy and water use (Gondhalekar and Ramsauer 2017).

Resilience is part of the language of urban climate adaptation. In this context, it refers to a city's capacity to absorb shocks and recover quickly. Climate change is a slow-onset hazard but its capacity to generate extreme weather outside of the usual conditions for which society has prepared will test urban infrastructure and institutions. Indeed, it already has in Houston, Miami, and Puerto Rico, not to mention Mumbai and Bangkok. In addition, weather-related events are not, by any means, the only shocks or stressors that cities will face in the future. Others include economic shocks, political shocks, terrorist attacks, and non-climate hazards such as earthquakes. Urban resilience links to institutional readiness, flexibility, and adaptability (Leichenko 2011). Thus, building capacity to deal with an uncertain climate may have benefits for cities facing uncertainty and surprise more generally.

New York and London were "early adapters" in the trend for urban climate adaptation. Rosenzweig and Solecki (2001) conducted a comprehensive assessment of 31 counties in the New York metropolitan area beginning in May 1998 with a charge from the United States Climate Research Program (USCRP) to develop a network of stakeholders to prepare an assessment of vulnerabilities and opportunities resulting from regional climate variability and change. The Metropolitan East Coast (MEC) study focused on seven sectors (coasts, infrastructure, wetlands, water supply, public health, energy demand, and institutional decision-making) to evaluate potential climate impacts through analysis of current conditions, past climate variability, scenario predictions, coping strategies, policy recommendations, and knowledge gaps. Later in 2010, under the leadership of Mayor Michael Bloomberg, New York developed a menu of 250 climate adaptation initiatives and put price tags on them. Three principles of New York's climate planning serve as exemplars for adaptation planning elsewhere. First, there is enough information to act on climate change today. In other words, it is possible to move forward in the face of uncertainty. Second, there is a plan to act across metropolitan regions rather than end at city boundaries because transportation, water, and other infrastructure systems are metropolitan in scale. Third, the plan should bring diverse interest groups together to build a shared understanding of climate vulnerabilities and science needs (Rosenzweig and Solecki 2013).

London's adaptation efforts began in October 2002 with the launch of "London's Warming: The Impacts of Climate Change on London" (UKCIP 2002). The process evolved from one focused on climate (statistical downscaling, emission scenarios, urban heat stress, flood risk, water resources, air quality, weather-related risk to transport systems, etc.) to building capacity to deal with the unexpected. The early report was a climate change 101 for stakeholders, drawing attention to the widespread and significant consequences of climate change for the city's economy, health, infrastructure, and social well-being. Nine years later, Boris Johnson's Report, *Managing Risks and Increasing Resilience*, was a call to arms for the city to build capacity to deal with extreme weather and long-term climate change and to export talent and services to the rest of the world. The report calls for co-benefit strategies such as increasing tree cover and improving public spaces to provide shade and flood protection (Nickson et al. 2011). The three pillars are (1) retrofitting London's existing buildings for more efficient energy and water use, (2) greening London to reduce the impact of noise and air pollution, making the city more resilient to flooding and extreme weather, and (3) cleaner air for London to reduce carbon emissions and improve health and quality of life. Flexibility is key to the London plan. Because climate will keep changing through the next century, society's responses to it must also change. Adaptation is more a "journey" than a "destination" (Nickson et al. 2011, p. 12).

Subsequent efforts in cities of the Global North and Global South developed and implemented many adaptation plans, most with an eye toward resilience, flexibility, monitoring, intersectoral coordination, scenario planning, and co-benefits. Co-benefits are particularly appealing in the context of climatic uncertainty because they are robust to climate change and reframe adaptation from a challenging problem to an opportunity to create jobs and economic opportunity, restore green space, improve quality of life, and build resilience.

Rotterdam (the Netherlands) is below sea level on the delta of the rivers Rhine and Meuse. It has developed a plan (Rotterdam Climate Proof) to augment the current system of storm surge, including barriers, canals, dikes, lakes, sewers, and pumping stations, that protects it from inundation. The old system was a masterpiece of Dutch engineering but inflexible and complex (Rotterdam Climate Initiative Climate Proof 2015). The city's adaptation strategy calls for food-proofing buildings, constructing flood-proof public spaces and floating communities, and restoring the sponge function of the area enclosed by dikes by capturing rainwater where it falls and delaying drainage. Green infrastructure features green roofs and facades, less paving and more plantings in public squares, and neighborhood water squares and infiltration zones as part of the infrastructure. The idea is to make Rotterdam a more attractive and prosperous city and, in the process, make it more resilient to climate change.

This mix of physical and green infrastructure and use of co-benefit strategies characterizes many urban adaptation efforts. Adaptation has moved beyond climate impact assessments to include coordination across sectors (Halifax, Calgary), water conservation (Melbourne, Denver, Hong Kong), land use planning (New York, London), active monitoring (Singapore, New York), and flexible decision-making (Sydney, Denver, New York).

Adaptation planning in the Global South is challenged by a lack of resources, viable models of action, and the sheer magnitude of making rapidly growing cities more resilient and sustainable. In addition, adaptation requires a more radical mindset than has previously characterized the planning profession and city leaders. Early-adopter cities often emphasized the internal goals and objectives of a local champion or entrepreneur. Late adaptors responded to natural hazards, national regulations, and incentives from development banks and foundations. Also significant was the drive to acquire legitimacy by aligning with institutional norms of adaptation. Carmin et al. (2012) systematically reviewed adaptation efforts in Durban, South Africa, and Quito, Ecuador, to gain perspective on the challenges of adaption in the Global South. An early assessment showed that Durban in 2100 would experience higher temperatures, increased rainfall, longer periods of drought, increased threats from vector-based diseases, sea level rise, coastal erosion, species extinction, property damage, and reductions in tourism revenue (Naidu et al. 2006). The Environmental Management Project launched a range of schemes including an integrated assessment tool showing future rainfall, runoff, temperature, crop yields, health impacts and sea level rise, efforts to make the 2010 FIFA World Cup soccer event carbon neutral, and community-based projects to promote food security. Lack of resources, focus on emergency management, and lack of national support have limited Durban's adaptation efforts.

The Quito Strategy for Climate Change (EQCC), released in February 2008, is a framework for implementing a climate agenda. It calls for an annual planning review, vulnerability analysis, and adaptation in the water sector. It was developed by an inter-institutional commission representing the Metropolitan Office of the Environment, members from the municipal air and water corporations, and the Strategic Research Unit. The commission initiated a public consultation process to engage residents, particularly those in vulnerable and marginalized neighborhoods to inform the climate adaptation strategy. After release of the initial strategy document, Quito created a Climate Change Office within the Environmental Office and started a process of annual planning, review, and implementation of the climate agenda. Quito did not develop a comprehensive adaptation plan with visions and guiding principles such as those in New York or London. Instead, they looked to ways to link adaptation to existing environmental planning. There, climate adaptation is an incremental process to reduce vulnerabilities in high-risk areas. Programs in both cities built on local resources rather than on a national or international template for adaptation, demonstrating the importance of place-based adaptation informed by the experiences and needs of local actors.

7.3 New Water (NEWater) in Singapore

Singapore is a small island nation with 5.5 million residents in the South China Sea off the coast of Malaysia. It is a model of urban climate adaptation and focused on water-related hazards and solutions (Fig. 7.1). Despite high levels of annual precipitation (2327 mm) (Meteorological Service Singapore 2017), Singapore suffers from water scarcity due to limited storage capacity related to its small land area and lack of groundwater (Bhullar 2013). From its inception as an independent country, it has been heavily dependent on nearby Malaysia for water supply. In a 1962 treaty enshrined in its constitution, Singapore is entitled to draw 1.14 million m^3/day from the Johor River in Malaysia, but this agreement is due to expire in 2061, and Malaysia has threatened to use water as a political tool (Williams and Liu 2015). Difficulties in solidifying an agreement to continue the relationship with Malaysia and the latter's willingness to use the water agreement as a political tool prompted Singapore to develop alternative sources to achieve water security and engage its population in novel ways to support water conservation and infrastructure.

Fig. 7.1 Singapore is a small island nation off the coast of Malaysia in the South China Sea

Singapore's adaptation effort takes the form of the "Four National Taps" strategy. The Taps include water from local catchments, imported water, reclaimed water (NEWater), and desalinated water. Some 65% of total demand now comes from reclaimed water and desalinated water; the remaining 35% originates in local catchments or is imported water. As part of its climate adaptation efforts, Singapore's Public Utilities Board aims to lower dependency on local catchments and imported water by increasing the production of reclaimed water to up to 50% by 2030 and 55% by 2060 (Public Utilities Board Singapore 2017a).

Singaporeans accepted reclaimed and desalinated water, in part, because of a strong and effective public education campaign (Tortajada and Joshi 2013). NEWater or reclaimed water is recycled/treated wastewater (sewage or used water) that undergoes conventional water treatment processes as well as stringent purification and treatment processes using advanced dual-membrane (microfiltration and reverse osmosis) and ultraviolet technologies. NEWater launched in 2003 and now supplies water from five plants and can meet up to 40% of the country's water demand. In addition, managers use NEWater in dry months to top up reservoirs and blend with freshwater at the waterworks before supplying to the population (Fig. 7.2). In order to increase the share of NEWater to meet 50% of demand by 2030 and 55% of demand by 2060, the government has proposed to increase the capacity of one of its plants and construct an additional NEWater plant (Public Utilities Board Singapore 2017a).

Acceptance of this very technical approach to water supply rests on public trust in the water treatment processes. One of the NEWater Plants is open for public tours, there is a visitor center, and the facility regularly hosts field trips by local students. The process of water purification is a source of national pride and symbol of Singapore's technological achievements. The NEWater Visitor Center includes not only an explanation of the technology but also a water museum to display the country's journey toward water sustainability.

Per capita domestic water consumption fell from 165 liters per day in 2003 to 151 in 2015, with plans to reduce further to 147 in 2020 and 140 by 2030 (Public Utilities Board 2017a). Public education ensures efficient use of existing water supplies. The country's policies require the

Fig. 7.2 NEWater facility in Singapore supports public understanding of the technology used to recycle water

installation of water-saving devices such as flow regulators, delayed action self-closing taps, and low-capacity flushing cisterns to conserve water. In order to maintain good water quality, the National Environment Agency requires water suppliers to implement and submit water safety and sampling plans for water production and plans. In addition, a recent policy mandate requires that licensed personnel perform all plumbing work to guarantee quality assurance for water customers (Public Utilities Board 2017b).

Efforts to implement conservation programs and build public support for investment in NEWater infrastructure leverage the country's image of a high-tech economy and island society. The city has transformed some of its water infrastructure into active, lively, meaningful public spaces. The Marina Barrage near the downtown core reduced lowland flooding in older neighborhoods. When it rains heavily during low tide, the barrage, which separates the reservoir from the sea, lowers its crest gates to release excess storm water from the reservoir. If heavy rains fall during high tide, the gates stay closed and large drainage pumps transport excess storm water out to sea. The barrage and related infrastructure functions are a public space in addition to flood control infrastructure. It is possible for people to canoe and boat in the Marina Basin, stroll across the bridge connecting the east and west sides of the Marina, and take in skyline views of the city. The Marina Bridge is an active public space and tourist attraction with a visitors center that explains how the water system works. Water features and play areas let visitors get close to water. Through play and interaction, the community develops a stronger bond with water and deeper understanding of its importance in sustaining lifestyles and the economy. Highlights of the new visitor's center include experience in what it is like to be a water molecule undergoing treatment, explanations of the membrane and ultraviolet technologies, and a range of education activities for local children and international visitors.

7.4 Metrics and Indicator Systems

Data, metrics, ongoing assessment, and evaluation are key components of DMUU and urban adaptation efforts. The need to respond to uncertainty requires information about how urban systems are evolving. The IPCC itself is an ongoing effort to assess available scientific and socioeconomic information about climate change and its impacts and options for mitigating climate change and adapting to it. Scientific assessments have consistently shown the extent of environmental change including glacial melting, shifts in geographic ranges of plants and animal species, and changes in biology that are consistent with a warming climate. More difficult is the monitoring of vulnerabilities and society's capacity to deal with them. The most recent IPCC report on urban areas notes that much of the emerging global risk is concentrated in urban areas, particularly those in middle- and lower-income countries. Urban vulnerabilities are increasing across the world in cities of all sizes, economic statuses, and geographic locations.

Urban indicator systems often fall under the umbrella of sustainability indicators. They enhance uncertainty management by allowing planners who can theoretically assess the impacts of designs, infrastructure, and policies, pinpoint areas for future analysis, bring citizens into planning and adaptation efforts, and identify problem areas that deserve further attention.

Many cities also now use indicator systems to gauge programs and evaluate themselves relative to peers. Indicator systems tend to be locally relevant as adaptation issues are highly place-specific. The European Commission recently published a report on indicator systems for sustainable cities, stressing the need to emphasize bottom-up approaches to ensure local relevance and public engagement. Finding generalizable indicators for the effectiveness and sustainability of urban water systems has proven difficult, however, because of very different perspectives on how to value aspects of water systems such as technical performance, economic efficiency, efficiencies of surrounding systems of energy and food production, access and availability, and taste and color (Lundin 1999). These value-based differences hamper inter-city comparisons but speak to the larger need to integrate local views into water management.

Paterson et al. (2015) argue for the potential of footprint ideas (Chap. 4) as a way of generalizing and comparing water use for cities. A city's water footprint has much the same meaning as a country's or individual's water footprint—how much water flows into and out of city boundaries through the cross-boundary exchange in goods and services. Instead of focusing on the expansion of infrastructure development as a vehicle for augmenting urban water supplies, the question changes to what are the implications of virtual flows for planning and policy. Water footprint studies reveal that Berlin obtains a majority of its virtual water supplies from outside of Germany, while water supplies for Delhi and Lagos come primarily from local sources due to differences in dietary patterns (Berliners eat more meat and drink more coffee).

7.5 Conclusions

Urban adaptation to climate change is a hotbed of DMUU thinking about building resilience to uncertain water futures. New York, London, Singapore, and Rotterdam have made significant progress in adapting water and related sectors to climatic and other uncertainties. Other cities have made indirect progress in water via new standards for land planning and efforts to protect vital energy infrastructure. Singapore is unique in its focus on water as the centerpiece of adaptation. Examples abound, however, to show that water is a potential threat to urban economies, environments, and vulnerable populations.

From its beginning, the urban adaptation movement has stressed citizen participation to build consensus for action and a shared understanding of the problem at hand. Lessons learned have furthered understanding of how science, decision-makers, and citizens can collaborate to build resilience through social learning. Chapter 8 articulates those lessons learned.

References

Bhullar Loveland. 2013. "Climate Change Adaptation and Water Policy: Lessons From Singapore." *Sustainable Development* 21(3): 152–159.

Carmin, JoAnn, Isabelle Anguelovski, and Debra Roberts. 2012. "Urban Climate Adaptation in the Global South: Planning in an Emerging Policy Domain." *Journal of Planning Education and Research* 32(1): 18–32.

Gill, Susannah E., John F. Handley, A. Roland Ennos, and Stephan Pauleit. 2007 "Adapting Cities for Climate Change: The Role of the Green Infrastructure." *Built Environment* 33(1): 115–133.

Gober, Patricia, Ariane Middel, Anthony Brazel, Soe Myint, Heejun Chang, Jiunn-Der Duh, and Lily House-Peters. 2012. "Tradeoffs between Water Conservation and Temperature Amelioration in Phoenix and Portland: Implications for Urban Sustainability." *Urban Geography* 33(7): 1020–1054.

Gondhalekar, Daphne and Thomas Ramsauer. 2017. "Nexus City: Operationalizing the Urban Water-Energy-Food Nexus for Climate Change Adaptation in Munich, Germany." *Urban Climate* 10(January): 28–40.

Hunt, Alistair and Paul Watkiss. 2011. "Climate Change Impacts and Adaptation in Cities: A Review of the Literature." *Climatic Change* 104: 13–49.

Kirshen, Paul, Matthias Ruth, and William Anderson. 2008. "Interdependencies of Urban Climate Change Impacts and Adaptation Strategies: A Case Study of Metropolitan Boston USA." *Climatic Change*: 86(1–2): 105–122.

Leichenko, Robin. 2011. "Climate Change and Urban Resilience." *Current Opinion in Environmental Sustainability* 3: 164–168.

Lundin, Margareta. 1999. "Assessment of the Sustainability of Urban Water Systems." Department of Technical Environmental Planning, Chambers University of Technology, Göteborg, Sweden. http://www.cpm.chalmers.se/CPMDatabase/DataReferences/TEP_1999--7.pdf. Accessed August 17 2017.

Mansell, M. G. 2003. *Rural and Urban Hydology*. London: Thomas Telford.

Maunsell Australia Pty Limited and CSIRO. 2008. "Impact of Climate Change on Infrastructure in Australia and CGE Model Inputs." Prepared by Maunsell Australia Pty Limited in association with CSIRO Sustainable Ecosystems. http://www.garnautreview.org.au/CA25734E0016A131/WebObj/02-AInfrastructure/$File/02-A%20Infrastructure.pdf. Accessed September 18 2017.

Meteorological Service Singapore. 2017. "Climate of Singapore." http://www.weather.gov.sg/climate-climate-of-singapore/. Accessed August 17 2017.

Naidu, Shamini, Rob Hounsome, and Kogi Iyer. 2006. "Climatic Future for Durban: Revised Report." Prepared for EThekwini Municipality and CSIR Environmentek. http://www.assaf.org.za/files/2010/LCC/Climatic_Future_for_Durban_Final_Report.pdf. Accessed September 18 2017.

Nickson, Alex, Helen Woolston, Juliette Daniels, Isabel Dedring, Kevin Reid, Kulveer Ranger, Louise Clancy, Roger Street, Tim Reeder. 2011. "*The London Climate Change Strategy*." London: Greater London Authority. https://www.london.gov.uk/sites/default/files/gla_migrate_files_destination/Adaptation-oct11.pdf. Accessed June 24 2017.

Oke, Timothy R. 1982. "The Energetic Basis of the Urban Heat Island." *Quarterly Journal Royal Meteorological Society* 108(455): 1–24.

Parry, M. L., O. F. Canziani, J. P. Palutikof, P. J. van der Linden, and E. E. Hanson (eds). 2007. "Increasing Human Utilisation of the Coastal Zone." Contribution of Working Group II to the Fourth Assessment Report of the Intergovernmental Panel on Climate change. Cambridge, UK and New York: Cambridge University Press. https://www.ipcc.ch/publications_and_data/ar4/wg2/en/ch6s6-2-2.html. Accessed August 16 2017.

Paterson, Willa, Richard Rushforth, Benjamin L. Ruddell, Megan Konar, Ikechukwu C. Ahams, Jorge Gironás, Ana Mijic, and Alfonso Mejia. 2015. "Water Footprint of Cities: A Review and Suggestions for Future Research." *Sustainability* 7: 8461–8490. https://doi.org/10.3390/su7078461.

Population Reference Bureau. 2017. "2017 World Population Data Sheet." http://www.prb.org/pdf17/2017_World_Population.pdf. Accessed August 14 2017.

Public Utilities Board Singapore. 2017a. "Singapore Water Story." https://www.pub.gov.sg/watersupply/singaporewaterstory. Accessed September 28 2017.
Public Utilities Board Singapore. 2017b. "PUB Revises Licensing for Plumbers from 1 April 2018." https://www.facebook.com/notes/pub-singapores-national-water-agency/media-release-pub-revises-licensing-for-plumbers-from-1-april-2018/1300201450070533/. Accessed September 28 2017.
Rosenzweig, Cynthia and William D. Solecki. 2001. "Climate Change and a Global City Learning from New York." *Environment: Science and Policy for Sustainable Development* 43 (3): 8–18.
Rosenzweig, Cynthia, and William Solecki (eds.), 2013. *Climate Risk Information 2013: Observations, Climate Change Projections, and Maps*. New York City Panel on Climate Change. http://www.nyc.gov/html/planyc2030/downloads/pdf/npcc_climate_risk_information_2013_report.pdf. Accessed August 17 2017.
Rotterdam Climate Initiative Climate Proof. 2015. "Rotterdam Climate Change Adaptation Strategy." http://www.rotterdamclimateinitiative.nl/documents/2015-en-ouder/Documenten/20121210_RAS_EN_lr_versie_4.pdf. Accessed June 24 2017.
Tortajada, Cecilia, and Yugal K. Joshi. 2013. "Water Demand Management in Singapore: Involving the Public." *Water Resources Management 27*(8): 2729–2746.
United Kingdom Climate Impact Program (UKCIP). 2002. "London's Warming: The Impacts of Climate Change on London." Summary Report. http://climatelondon.org.uk/wp-content/uploads/2012/10/Londons-warming-summary.pdf. Accessed June 24 2017.
Williams, Jessica and Su Liu. 2015. "The Water Tales of Hong Kong and Singapore: Divergent Approaches to Water Security." Technical Report. https://doi.org/10.13140/RG.2.1.3009.2643. Accessed September 18 2017. Accessed September 18 2017.

8

Social Learning for Water Sector Resilience

8.1 Introduction

Water resource modeling supports a wide range of water management decisions, including issues of water availability, quality, and variability. Model results inform decisions about infrastructure construction and operations, flood control and drought management, harvesting and storing water above and below ground, healthy ecosystems, and the allocation of water toward crops, industries, cities, and environmental uses. Models can lead to scientific discovery such as addressing the anticipated impacts of climate change on regional hydrological systems (Cook et al. 2015) or the hydrologic dimensions of climate-soil-vegetation dynamics (Rodriguez-Iturbe 2000). They are also relevant to public interest in and social support for adaptive management and collective action (Pahl-Wostl et al. 2013; White et al. 2015). In the latter, model building is an ongoing process of social learning rather than a vehicle for scientific prediction or one-off decision support.

Resilient water systems function at the triple junction of scientific discovery, decision support, and public engagement (Fig. 8.1). New scientific knowledge about non-stationary systems (including biophysical and

P. Gober, *Building Resilience for Uncertain Water Futures*,
https://doi.org/10.1007/978-3-319-71234-5_8

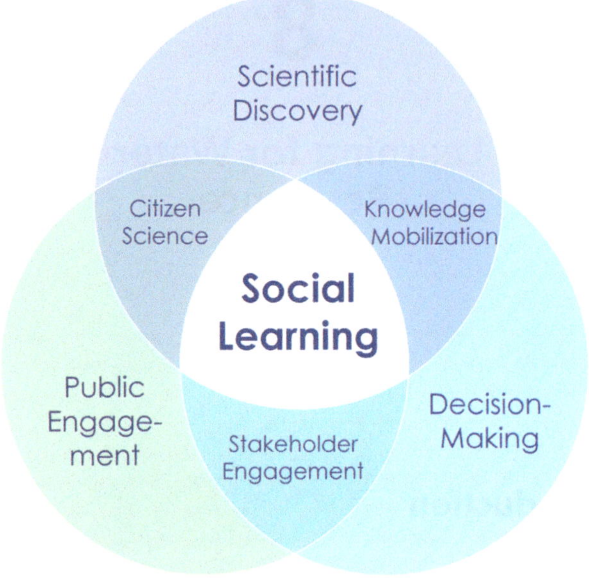

Fig. 8.1 Social activities surrounding water resource modeling

societal uncertainties) informs management decision-making and facilitates citizen engagement with science. Citizen science builds capacity into scientific institutions and organizations for knowledge that is inclusive, accountable, evolving, and reflective of diverse viewpoints. Stakeholder engagement with the decision-making community ensures that management decisions are both evidence-based and reflective of human beliefs, attitudes, and values (Glynn et al. 2017). Social learning, a process by which science, decision-making, and public engagement intersect, involves the joint capacity to learn about and respond to ongoing water system dynamics. Managing water problems increasingly requires the ability to administer complex, dynamic, and uncertain water systems and the controversies that surround them. The social infrastructure that supports water system development is critical to resilience—the capacity to cope with new information, new technologies, and new climate conditions.

This chapter lays out the social history of water science, engineering, water infrastructure construction, and human relationships with water. It provides the context for needed changes in the water sector. It then articulates why social learning is a critical ingredient of water-sector resilience, provides examples of social learning experiments, and discusses best practices in knowledge mobilization, citizen science, and stakeholder engagement. The idea of knowledge mobilization involves translating scientific knowledge and models into tools for decision-making. Citizen science addresses the need for public support for science-based water issues and enables scientists to learn from water users in the community. Stakeholder engagement ensures that local viewpoints inform water decisions. Indigenous peoples see themselves as self-determining nations that pre-date colonial contact. As such, their engagement in collaborative exercises of water governance should reflect their special status as rights holders rather than as one of many groups with a stake in water decisions (von der Porten 2014).

8.2 History of Water and Society

The need for social learning stems in part from the physical separation of people from the water systems that sustain their lives and the economy, the historical disconnection of water science from its user communities, and the many uncertainties and controversies that characterize modern water management. For most of human prehistory, hunters and gatherers consumed plants and animals that depended on rainfall that varied significantly from place to place. The Agricultural Revolution, associated with the domestication of food production some 10,000 years ago, led to the concentration of population around water sources such as rivers and groundwater and the rise of irrigation, populated settlements, specialization and the division of labor, economic trade, and complex social organizations (Hassan 2011). People lived close to water sources, and used these spaces for ceremonial and spiritual purposes along with economic and practical uses.

Nineteenth-century industrialization brought population from rural to urban areas. Water development moved toward the widespread use of irrigation and drainage methods, long-distance water transport, and large-scale storage. Cities built pumping stations, water towers, dams, reservoirs, and sewage treatment plants to improve water safety and reliability (Hassan 2011). This infrastructure began to separate people from the water sources that sustained their lives.

A good example of this is in Winiwarter et al.'s (2016) study of nineteenth-century water development in four European riverine cities: Vienna, Lyon, Munich, and Brussels. Prior to large-scale industrialization, the cities differed in land area, population, natural and cultural endowments, administrative structure, and power relations. All grew rapidly after 1850. Drivers of change in water systems included the need for river engineering and flood protection, advances in the fight against cholera, and demand for improved sanitation, water supply, and sewage networks. By the end of the nineteenth century, the four cities had water networks that were more alike in structure and function. They followed similar paths of water development despite differences in human and physical geography. Small watercourses became part of the subterranean sewage system, and large ones were divorced from their floodplains by engineered banks to provide flood protection. City governments delegated management of water systems to a new professional class of engineering experts. The modern water network disconnected people from water and damaged riparian ecosystems. The authors warn that city dwellers today use large amounts of fossil fuels to maintain their vast underground networks of water infrastructures and experience vulnerabilities from the significant public investments required to maintain them. Current residents have in their words "accepted a bill whose footing has yet to come" (Winiwarter et al. 2016, p. 230).

The hydraulic mission of controlling water systems dominated North American and European development after the 1830s (Anderson 2004). Civil engineers tackled problems of waterway improvements, coastal engineering, irrigation, hydropower generation, the design of urban water supply and treatment systems, and flood and drought management. In 1802, the US Congress created the Army Corps of Engineers to build and fortify military installations. The Corps' mission expanded in 1824 to map the new country and serve as the nation's water manager in terms of navigation and flood control. Congress appropriated funds for improving navigation on the Mississippi and Ohio Rivers and authorized the Corps' engineers to use their expertise to determine the most practical way of doing so (US Army Corps of Engineers 2017). Dutch water management was a "hydraulic technocracy" in which engineers had discretion in identifying problems and potential solutions without non-expert consultation (Disco and Toussaint 2014). Plans related to the two largest coastal projects in the Netherlands between 1900 and 1970 were developed from studies and recommendations prepared by engineers based on exhaustive research. For much of its modern history, the field of water engineering was about control of water resources, putting them to use in support of economic development and human needs, and investing the profession with discretion to solve water resource problems from a technological perspective.

Big dams and the reservoirs behind them are signature developments of twentieth-century water engineering. They allowed society to manage the variability in water supply and demand and serve other important purposes such as flood control, recreation, and navigation. The Global Reservoir and Dam Database shows the early concentration of dams in North America, Japan, and Europe and more recent development in China, Latin America, and Africa (Figs. 8.2 and 8.3 and Table 8.1). Dam construction worldwide peaked in the 1960s and 1970s when storage capacity grew annually by 150 km^3 in 130 new reservoirs. This compares to annual storage capacity growth of only 15 km^3 between 2001 and 2010 (Wisser et al. 2013). Sedimentation has reduced the expected life span of many dams, suggesting that we may be beyond peak reservoir capacity. Sedimentation loss rates vary from 0.5 to 1% annually with large variations depending upon site and locational characteristics (Mahmood 1987; White 2001; Wisser et al. 2013). Some of the large North American dams designed for 150 years of use have lost 50–100 years of potential use due to sedimentation (Hargrove et al. 2010). In terms of climate change, this means societies are losing capacity to buffer uncertainties in supply and demand through water storage. This declining capacity differs significantly from one river basin to another.

Critics raised questions about the sustainability and environmental costs of large dams in the 1960s and 1970s. In addition to capacity loss due to sedimentation, other warning signs emerged in the form of acute pollution, population growth and increasing use, overdevelopment of river basins, large numbers of people displaced by dam construction,

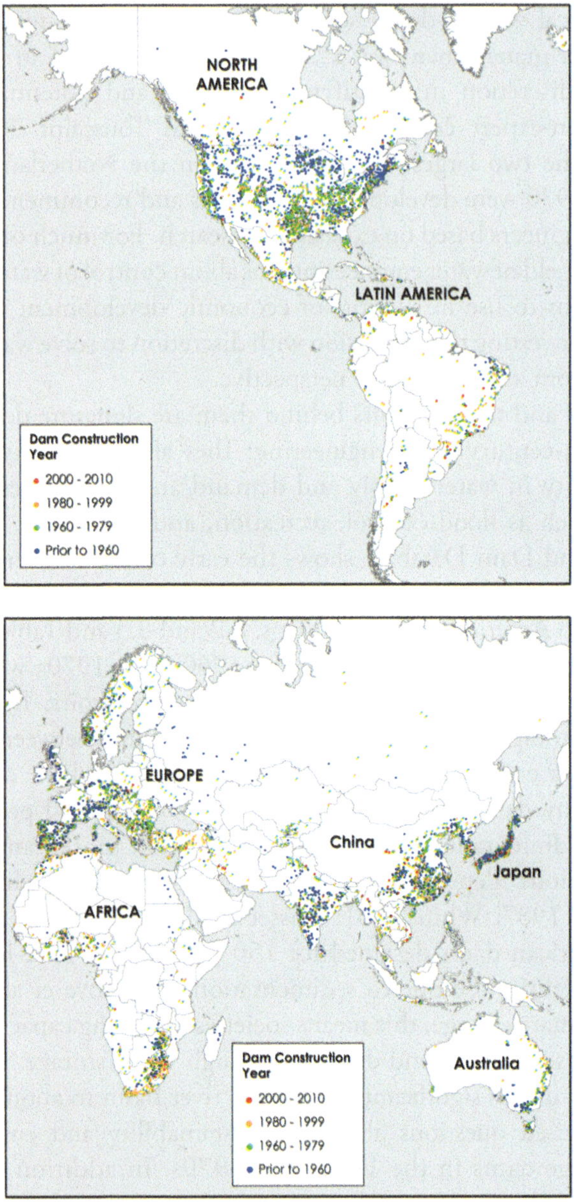

Fig. 8.2 Modern dam development was concentrated in North America and Europe and more recently in China. Global Reservoir and Dam (GRanD) Database

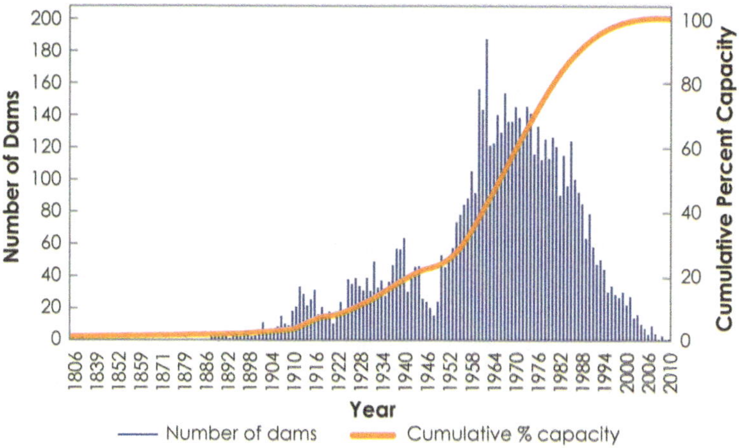

Fig. 8.3 Global dam construction, 1800–2010: (a) number of dams and (b) cumulative reservoir capacity. Global Reservoir and Dam (GRanD) Database, 2011

Table 8.1 Dam development in North America, Europe, China, Japan, and the rest of the world

Region	Number of dams	Average year of construction	Average max. capacity	Standard dev. of max. capacity
North America	2196	1949	345.76	1613.26
Europe	1304	1964	40.98	272.75
China	611	1970	113.47	1830.37
Japan	542	1962	97.45	15.21
Rest of World	1798	1964	167.27	2817.81

Source: Global Reservoir and Dam Database. 2011. "GRanD Database." Accessed August 18, 2017. http://www.gwsp.org/products/grand-database/global-reservoir-and-dam-grand-database-project.html

degradation of aquatic ecosystems, and loss of biodiversity (Molle 2009; Vörösmarty et al. 2010). Management in Europe and North America began to move away from infrastructure construction to greater emphasis on sustainable management of water resources, including the impacts of water development on ecosystem health and opportunities for green infrastructure.

Dam construction persisted in China as a means to deal with water scarcity and as a vehicle for economic development (Fig. 8.4). China

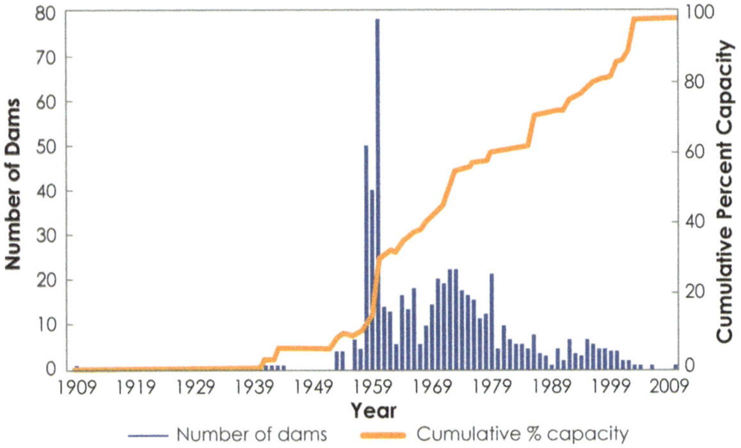

Fig. 8.4 Dam construction in China 1909–2009: **(a)** number of dams and **(b)** cumulative reservoir capacity. Global Reservoir and Dam (GRanD) Database

built many small- and mid-sized dams between 1950 and 1980, followed by fast growth in capacity via the construction of large dams from 1980 to the present. Note the very large standard deviation in maximum dam capacity that reflects the recent trend toward large-scale development in China (Table 8.1). The maximum storage capacity continues to rise in the twenty-first century. The Three Gorges Dam (TGD) built on the Yangtze River in 2003 is China's largest water transfer project with 39.3 billion m^3 of storage (Global Reservoir and Dam Database 2011). Benefits of large dam construction are flood mitigation, water scarcity reduction, clean energy generation, and increased water for food production, but the negative consequences have also been significant. They include downstream channel erosion, loss to sediment discharge to the delta, loss of biodiversity, and other ecological changes in lower reaches. Dam construction in China has led to the relocation of some 22 million people (Chang et al. 2010; Liu et al. 2013). Today, China faces the challenge of balancing economic progress with ecosystem services and the dilemma of "keeping the flood away" or "giving the flood way" (Yin and Li 2001; Opperman et al. 2009).

New ideas about infrastructure and environmental protection are transitioning the field of water engineering away from trying to control nature to a more eco-centric and community-based approach to management (Molle 2009). Green infrastructure, first introduced in the fields of urban planning and landscape ecology, is one manifestation of this shift (Wade and McLean 2014). Traditional risk reduction strategies funnel rainwater through gutters, storm sewers, and engineered collection systems. Water quality degrades as runoff carries trash, pollutants, and heavy metals across the urban landscape. Green infrastructure seeks to reestablish natural drainage processes and allow natural systems to soak up, store, and cleanse urban water. Green infrastructure uses downspout disconnections, rainwater harvesting, rain gardens, planter boxes, permeable pavements, green streets and alleys, green parking lots, green roofs, and land conservation to slow drainage (US Environmental Protection Agency 2017). It provides opportunities for land and water managers to coordinate actions and for urban residents to reunite with the water-based landscapes and become more involved with water management at small neighborhood and home-based scales.

As water engineering evolved to meet the technical needs of modern societies, the scientific field of hydrology focused on the biophysical aspects of process dynamics. Disciplinary culture favored narrow analytical research questions at the expense of a more holistic, synthetic view of the water system. Sivapalan et al. (2012) recently called for a new science of "socio-hydrology," making the point that it is not possible to predict long-term change in water systems dynamics without considering the interactions and feedbacks among natural and human components of the water system. Feedbacks can move a system beyond critical tipping points into a new, previously unobserved state (the new normal).

Missing from this broader hydrological perspective are the social processes whereby scientific knowledge about hydrological and socio-hydrological systems links to decision-makers and the public. These two-way knowledge transfer processes are increasingly important in linking scientific knowledge and modeling (physical, social, and socio-hydrological) to decision-making and public attitudes and values. Wicked water problems require input from a variety of actors, including, but not limited to, water scientists. Solving complex, uncertain, and contentious water problems requires collaboration with decision-makers and the public at large in a process known as social learning.

8.3 Social Learning

Social learning is simply the idea that people learn as part of a group. Social learning theory, as set forth in the 1960s and 1970s, asserts that learning is a cognitive process that takes place in a social context. Learning occurs from observing behaviors and their consequences. The learner is not a passive recipient of knowledge. Cognition, environment, and behavior mutually affect each other (Bandura and Walters, 1963). Proponents of adaptive management of complex environmental systems use these ideas to construct a social infrastructure to support resilience to abrupt and uncertain change. Managing environmental resources for resilience requires, they say, the capacity to learn from experience, synthesize different types of knowledge and experiences, and view policies as learning experiments (Folke et al. 2005).

Ecologists led the way to integrate social learning and resilience into ecosystem management, but the idea has been applied in other fields such as urban planning (resilient cities), psychology (an individual's ability to adjust to life in the face of adversity), and natural hazards (the ability to bounce back after a natural disaster) (Adger et al. 2005). Enhancing the capacity to stabilize a complex system experiencing shock requires institutions and groups of individuals to respond quickly, learn from experience, and adjust to ever-changing non-stationary conditions.

Social learning and knowledge acquisition are an ongoing part of transition management—the transition from stationarity to non-stationarity. Expert knowledge combined with local knowledge can make sense of long- and short-term change and build trust and legitimacy for collaboration and collective action. As applied to decision-making, social learning builds capacity for teams to process information, make sense of scientific data, connect it to a particular context, and build organizations and institutions that are attuned to the need for continuous monitoring, self-reflection, and ongoing adaptation.

Social learning promotes a shared understanding of the challenges of climate change and shapes the political processes surrounding decision-making. It triggers reciprocity among different actors with an interest in the problem, but implies new roles for water scientists as participants in a larger societal dialogue rather than as freestanding actors delivering the quintessential evidence base for decision-making (Pahl-Wostl 2009; Pahl-Wostl et al. 2002, 2007, 2008, 2010, 2013). Hurricanes Harvey and Irma in the Southeastern US are unlikely to be one-off events in a changing climate but flood events that open windows of opportunity for policy action and offer new insights into disaster management and flood mitigation. Social learning provides a framework to build public support for collective action to reduce vulnerabilities that are at the heart of flood risk.

8.3.1 Knowledge Mobilization

Knowledge mobilization refers to moving knowledge from formal research projects into active use. It involves the sharing of knowledge between research scientists and decision-makers, often through the help of third parties. A US National Academy of Sciences Report, commenting on the status of climate change science, declared in 2007 that "inadequate progress has been made in synthesizing research results, assessing impacts on human systems, or providing knowledge to support decision making and risk analysis" (National Research Council 2007, p. 34). Reasons included a lack of meaningful interaction between scientists and decision-makers, organizational culture and reward structures, risk averse and vulnerable cultural contexts, hard to interpret presentation of scientific information, and user difficulty in translating probability information into recommendations for action. Potential remedies involve having users at the discussion table to frame the problem initially. Also significant is recognizing the importance of two-way, iterative engagement between producers and users to build trust and better understand the needs of policy-making (Dilling and Lemos 2011). Pahl-Wostl and Borowski (2007) noted similar problems in implementing the requirement for participatory water management as required by the European Water Framework Directive. Despite claims of usability and problem solving by scientists and the European Commission, many of the tools designed for decision support did not meet user needs.

Difficulties in producing usable climate science for policy-making devolve in part from the scientific enterprise itself (Dilling and Lemos

2011). Scientists often make assumptions about what they think will be usable that do not correspond to what users think is useful. The bands of uncertainty surrounding model projections are often of less interest to users than the worst-case outcome. Successful use of climate knowledge involves continuous iteration between the producers and users of knowledge. A common problem is that nobody owns the task of producing usable scientific information and takes responsibility for connecting scientists and users. The presence of intermediaries or knowledge brokers helps to ensure that the right kind of scientific information matches the right user. Boundary organizations have emerged to work at the interface of science and policy and ensure that mechanisms are in place that foster iteration, rewards for synthetic science, and capacity to hire new people committed to innovation in the mobilization of knowledge. A recently funded project (Global Water Futures) from the Canadian government for climate and flood modeling and decision support has a designated four-person knowledge mobilization team to support collaboration between scientists and users, ensure there is regular feedback between science teams and participatory user groups, and prepare team-based knowledge mobilization plans. Collaboration does not happen naturally or by accident; it needs to be an expected function of project management, like budgetary reviews, performance assessments, or scholarly output, in the case of science projects.

The US National Science Foundation funded the Decision Center for a Desert City (DCDC) in 2004 as a boundary organization to facilitate knowledge mobilization between university scientists and local water managers. It holds monthly Water-Climate Briefings to discuss a broad range of issues such as sustainable water reuse, urban heat island mitigation, political versus scientific uncertainty, the water-energy nexus, ecosystem planning, Native American water rights, green infrastructure, communicating complexity, and water development in Latin America. These events attract a mix of scientists, students, and local and regional water managers.

DCDC developed the exploratory model WaterSim explicitly to reach out to water managers and enhance social learning. The model progressed in multiple iterations from WaterSim 1 to WaterSim 5 to capture issues raised by local stakeholders, such as water reuse, land use, water rights, and the need for provider-level analysis. DCDC visualized model results and used them for scenario and policy discussion in an immersive Decision Theater (Figs. 8.5 and 8.6). WaterSim was intended to fit Cash et al.'s (2003) notion of a "boundary object," developed jointly by scientists and users and capable of maintaining an identity across them. Boundary objects are more likely to produce salient knowledge because users are engaged early in the development process, credible knowledge because they bring multiple types of expertise to the table, and legitimate knowledge by providing stakeholders with access to the model building process.

Empirical studies of knowledge mobilization activities show that water stakeholders often begin the science engagement process with very different priorities and concerns about water problems. In an early study at DCDC of how stakeholders, including representatives of federal agencies, Indian tribes, state entities, local water departments and regional agencies, regional public water providers, private water providers, private sector users, and environmental groups, defined sustainable water and prioritized evaluation concerns, survey results showed substantial diversity across stakeholder groups. Representatives from local water departments and regional agencies emphasized the sufficiency of water supplies,

Fig. 8.5 WaterSim in the Decision Theater. Liz Marquez/Decision Center for a Desert City/Arizona State University

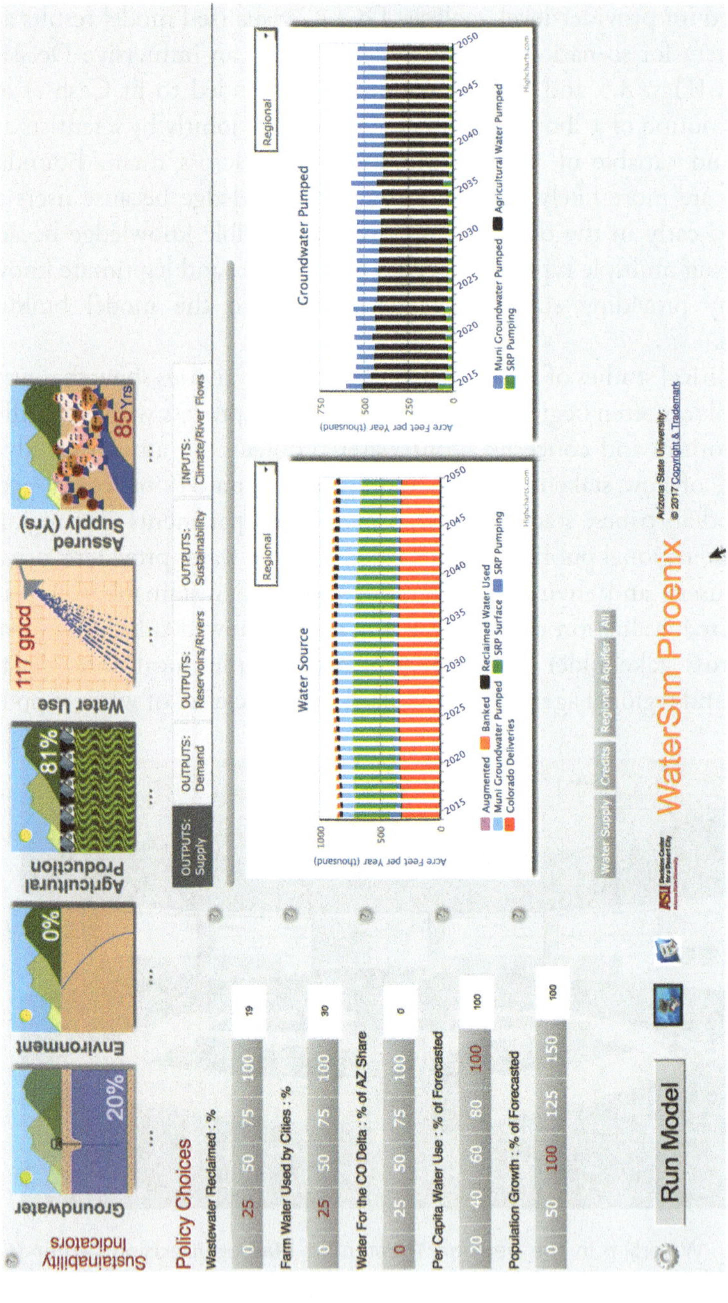

Fig. 8.6 Graphical interface for WaterSim 5. Liz Marquez/Decision Center for a Desert City/Arizona State University

whereas members of environmental groups placed high priority on ecosystem health. Results pointed to the importance of the trade-off between system reliability and environmental flows as a basis for public discussion, knowledge mobilization, and social learning in Central Arizona (Keller et al. 2010).

White et al. (2015) conducted ten stakeholder focus groups including water managers, consultants, and data analysts to evaluate WaterSim, specifically with respect to the issue of scientific versus political uncertainty. Scientific uncertainty involves margins of error around model predictions; political uncertainty entails the costs of being wrong in a political context. Decision-makers may be willing to accept a higher margin of scientific uncertainty if the political cost is low, but tolerate little risk if the political costs are high. Focus group results showed that participants framed uncertainty in both scientific and political terms. In other words, they were concerned about the capacity of WaterSim to make accurate predictions and about the political context in which model results are used. They were concerned about the assumptions and data, technical aspects of model construction, and the efficacy of the model results. They were also interested in the political context. There was a sense that model results could induce a sense of urgency into water decision-making, but also a concern that they would raise thorny political issues about growth management and climate change. Overall, stakeholders saw the model as an opportunity to narrow the gap between scientific and political uncertainty.

In Western Canada, water stakeholders involved in an early effort to support knowledge mobilization at the Global Institute for Water Security were asked to describe what "water security" meant to them and rank various water security issues (climate change, flooding, drought, water quality, water quantity, competing demands, governance, and land-use management) in terms of the priority of their concerns (Gober et al. 2015). A majority of respondents defined water security from a sustainable development perspective, mentioning future needs, the triple bottom line (economic development, social justice, and environmental health), and value-based trade-offs. Moving beyond a strictly sustainability perspective to include a more value-based definition, one First Nations respondent defined water security as being "safe, accessible, grounded, sacred, natural."

In response to a request to share their thoughts about barriers to water security in the region, respondents unleashed a torrent of complaints about poor governance, lack of leadership, over-allocation, lack of transparency in decision-making, and declining flows due to climate change impacts. Also in response to the barriers-to-security question, one respondent mentioned "greed by all of us, particularly oil, gas, and mineral extraction interests." Another blamed public sector leaders for "valuing economic growth over inclusive responsibility." Overall, stakeholders expressed far less concern about climate change impacts per se than about the human capacity to manage these impacts.

Results produced a cautionary message for scientists hoping that their climate and hydrological research would automatically inspire action to reduce climate change-related water vulnerabilities. When asked to prioritize their concerns about water, stakeholders expressed greater concern about the human over the physical dimensions of water security. Water quality ranked first, governance second, quantity third, land use management fourth, and competing demands fifth. Concerns about biophysical aspects of water management, including drought (sixth), long-term climate change (seventh), and flooding (eighth), were lower-priority concerns. The take-home message is that scientists will increasingly need to frame their research and outreach to capture the policy, governance, and economic concerns of water stakeholders if they are to produce usable science about water security.

Not all social learning/knowledge mobilization lessons come from formal surveys, interviews, and intentional activities. In collaboration with the Arizona State University (ASU) Decision Theater and the Arizona Water Institute, DCDC hosted a set of four collaborative workshops about uncertainty and climate modeling for climate modelers across Western North America and water managers in 2009–2010. Topics included integrated water planning, robust decision-making, downscaling climate models, and modeling uncertainty. The downscaling workshop paired local and regional water managers with global climate modeling teams. Discussion first centered on techniques for downscaling and managing uncertainties. It quickly moved into the technical details of parameterization with a discussion among the modelers of the different ways to manage uncertainty in the downscaling process. Water managers were excluded from this highly technical discussion, but in later feedback they reported that the disagreements among the scientists had given them new insight into the sources of uncertainty in climate model downscaling and why the models produced different results. These insights, although unintended, provided a useful context for evaluating models and the vulnerability analyses they produced.

DCDC spent more than a decade building capacity for knowledge mobilization between water managers and scientists. Joint activities ranged from workshops, monthly luncheons, and panel discussions, to joint research and an internship program. These activities allowed scientists to learn more about the institutional context for decision-making and the motivations that drive decision-makers. At the same time, policymakers learned more about the research process and the challenges of creating knowledge for decision support in a traditional research environment (Quay et al. 2013).

Box 8.1 Principles and Best Practices for Knowledge Mobilization

1. The research "team" consists of researchers and practitioners working together to co-produce knowledge. End users are known potential adopters or users of the information the team creates.
2. The research project design is for an end-user audience. To this end, scientists determine how the end-user community normally collects, evaluates, and uses knowledge.
3. There is a plan for knowledge mobilization at the outset. It is built into the research design, paying particular attention to the kinds of research outputs that end users need and can use (e.g., topic, format, language), and when the knowledge will be useful (e.g., during certain points in their planning cycles and outreach efforts).
4. Researchers determine how users typically communicate with each other, and then create opportunities for active, inclusive, and iterative communication between research team members and them, using users' preferred communication channels. Forums for sharing information (workshops, luncheons, presentations, webinars, formal written reports) are compatible with end user needs.
5. Mutual understanding within the team requires a shared vocabulary. The use of jargon and acronyms is limited. The research team takes responsibility for translating technical terms into plain English, as a matter of course.

6. Boundary objects, such as jointly created conceptual frameworks, data sets, and models, are effective tools for knowledge mobilization.
7. Boundary organizations take responsibility for convening scientists and stakeholders, mediating the boundary between scientific research and practice, and ensuring that participants work toward a shared understanding of the problem or opportunity. In an interdisciplinary team, these differences may not simply be between researchers and practitioners; they may also exist among researchers themselves.
8. Ideas and information flow in all directions within the team. Lines of accountability are clear between researchers and practitioners within the team, and between the team and any intermediaries who support knowledge mobilization.
9. Communities of practice take responsibility for sharing knowledge beyond the specific project team to enhance social learning outside the particular group of scientists and users.

8.3.2 Citizen Science

Science outreach also includes engagement with the wider public about important new research findings and the role of science in social learning and adaptive management. Rather than being neutral observers, scientists generate critical information about the changing state of biophysical systems, knowledge and understandings of complex system dynamics, and assessments of winners and losers from any particular policy decisions (think about the US Congressional Budget Office that summarizes the costs and benefits of policy decisions discussed by the US Congress). The policy evaluation process begins with the acquisition of data and then structures that data around local belief systems in a way that facilitates action (Glynn et al. 2017). Subsequent monitoring and evaluation updates existing conceptual models and alerts society of the need to adjust policy in response to changing signals from the environment and society. In this model, science does not produce the penultimate answer to a pre-determined policy question but rather participates in a DMUU discussion.

The Global Institute for Water Security at the University of Saskatchewan sponsored a theatrical performance, *Downstream*, to communicate water problems from a humanistic perspective to a diverse audience of practitioners, scientists, students, and members of the public (Strickert and Bradford 2015). *Downstream* involved a Forum-Theater production where audience participation is encouraged. It involved professional playwrights, actors, and costume and set design experts. The narrative conveyed the downstream consequences of a flood event involving an irrigated farm upstream, via fish in the river, a First Nations-owned golf course, a municipal water treatment plant, a Prairie town, and a mid-stream oil field to a Saskatchewan River Delta village downstream. At each point along the way, relevant characters pleaded with the audience for the resources needed for flood management, and deflected insults from other stakeholders. The audience allocated resources in response to the theatrical performance (Strickert and Bradford 2015). The interactive play conveyed the complex context for flood management decision-making, used science-based flood scenarios, and called upon the audience's values, emotions, and attitudes about flood risk for mitigation strategies, policy reform, and adjudicating among competing interests (Fig. 8.7). It linked emotions about water with water science via a structured activity and the social process of Forum-Theater.

Fig. 8.7 *Downstream*, a Forum-Theater performance. Graham Strickert/Global Institute for Water Security/University of Saskatchewan

The play unfolded as a game show where the different sectors competed with one another for audience support. Decision-makers received a bag of ping-pong balls and were asked to allocate their "resources" at critical junctures in the timeline. Multiple decision points represented the iterative nature of science–public interactions. Three hundred people participated in *Downstream* during its run in February 2014.

Adaptation in the water sector is less about achieving a particular end (e.g., 25% reduction in use or 100% reliability) and more about creating a set of processes that facilitate resilience (Pearson et al. 2010). Public engagement with science ensures acceptance, ownership, and buy-in to later decisions; it also facilitates the social learning processes that typically involve a variety of different types of knowledge (Pahl Wostl 2009). Public participation also creates opportunities for decision-makers to resolve contentious trade-offs and enhances the prospect that decisions hold across numerous political cycles.

Box 8.2 Principles and Best Practices in Citizen Science and Public Engagement

1. Knowledge about water systems comes from a variety of sources, including Indigenous, local, spiritual and cultural, artistic, and scientific. They offer complementary understandings of how water, land, and ecosystems change over time and the role of human intervention in complex systems.
2. Engage local actors in scientific research programs early and often to elicit salient research issues and identify social and political concerns. One-time activities are unlikely to be impactful. Sustained activities and institutionalized engagement are more likely to achieve the goal of linking local knowledge to water science issues.
3. Back-and-forth communications assume there is more than one way to understand a scientific issue. Scientists do not own the research problem, with public participants being passive consumers of scientific knowledge.
4. The public engagement process should involve different avenues of communication including in-depth workshops, deliberative forums, focus groups, public opinion surveys, and scenario-building exercises. Eliciting public participation is not a one-size-fits-all process.
5. Scientists should write for the public with an interdisciplinary, professional audience in mind. There should be a clear articulation of how research aims and products translate into broader impacts for society.

6. Science cafes are casual events that host conversations between scientists and the public about current science topics. They take place in coffee houses, bookstores, and restaurants and employ a moderator for public discussion of science issues.
7. Science festivals bring together temporary exhibits, museum-type activities, art organizations, students, and members of the public.
8. University-led cooperative engagement fosters trust building and social learning with stakeholder groups such as farmers, coastal landowners, minority groups, and industry partners.
9. Although hydrology historically had limited engagement with local citizens, big data, mobile apps, and low-cost sensors can engage citizens with water science. Low-cost measurement can expand sensing coverage and the intensity of rainfall, streamflow, and water quality sampling in much the same way that bird sightings have enhanced coverage in ecology.
10. Participants who show up for engagement events are more likely to be knowledgeable about the topic and feel intensely about it (e.g., flooding, drought, water quality, climate change). Extra effort is required to draw out participants who can reveal health and environmental justice concerns.

8.3.3 Stakeholder Engagement

Planner Sherry Arnstein's 1969 Ladder of Citizen Participation provides a framework for assessing the substance and authenticity of stakeholder participation in public decision-making (Arnstein 1969). Her original typology used participation in federal social programs of the 1960s such as urban renewal, anti-poverty, and model cities as prototypes for steps in the ladder. It has been applied to a wide variety of natural resource management issues, including water (Parkins and Mitchell 2005; Tippett et al. 2005; Akhmouch and Clavreul 2016). Arnstein differentiated citizen participation based on power relationships with decision-making agencies. At the bottom are manipulation and therapy, which are, in fact, forms of non-participation and tokenism. They allow those who hold power to "educate" participants without giving up control over the learning process. Moving up the ladder leads to the gradual shift from informing, consultation, and placation to partnership, delegated power, and citizen control in which citizens gain full management authority. Arnstein envisioned a social justice agenda, moving toward a redistribution of power in society allowing "have-nots" to join in, determining how information is shared, goals are set, resources are allocated, programs are operated, and benefits are distributed (Arnstein 1969).

Water decision-makers increasingly face choices about how to manage the resource in an equitable, efficient, and environmentally sound manner in the face of scarcity, uncertainty, and different value frameworks. Stakeholder engagement raises the prospect of improving acceptance and trust in water governance and reducing conflict over water issues. This is the logic in forming watershed steward groups, eliciting formal consultations about strategic planning, and inviting farmers to form drainage networks to protect downstream neighbors and wetlands from further development. The trend is to move away from one-off, top-down hierarchical methods of engagement to facilitate ongoing engagement from local actors in a more systematic way. Akhmouch and Clavreul (2016) advocate for moving from efforts to hold public hearings and meetings to more proactive efforts and targeted programs to share ideas about a specific problem such as drought management, flood hazard preparation and mitigation, or a particular water quality incident. Common barriers to participation are the lack of political will and leadership, lack of clarity about exactly how input relates to decision-making, institutional fragmentation with overlapping responsibilities, lack of funding for a sustaining engagement effort, and consultation fatigue.

Box 8.3 Principles and Best Practices in Stakeholder Engagement

1. Avoid techniques such as manipulation and tokenism. Emphasize partnership and power sharing. Successful engagement moves from early consultation to shared decision-making later in the process.
2. Activities fit within a policy decision window when the need for change is widely acknowledged and participants feel they can make a difference. Participant fatigue occurs when the same people fill the same roles without seeing meaningful progress toward policy change or problem solving.

3. The legitimacy and credibility of public engagement rests in part on the ability to assemble a representative cohort of the public. The views they voice should be a valid expression of public opinion. This can occur at one session or with multiple focus groups or dialogue sessions expressing competing viewpoints about the problem at hand.
4. There are multiple modes of engagement, depending upon stage in the decision process. Early events engage the public in exploratory conversations and broker the pros and cons of potential decisions, with participants raising social and ethical issues. Later events mobilize the public for action on agreed-upon policy.
5. Local users of water resources sometimes have little substantive information about relevant potential problems such as floods and droughts. Public participation can be a mechanism for good governance—to infuse public scrutiny of infrastructure and operational decisions that affect their lives.
6. Organizations investing in a developing country's water infrastructure should require evidence of consultation with parties affected by the decisions of their organizations.
7. Participation often reflects the role of gender access to water and sanitation services. Women and girls in developing countries are most often the primary users, providers, and managers of water in their households and are responsible for household hygiene. Equal participation from men and women enables multiple perspectives on access to water services.
8. Effective public participation requires the sharing of information. Data such as water use per capita are often difficult to obtain for comparative purposes, because providers often do not use standardized methods of data collection and organization. Data presentations should allow audiences to view this knowledge in a larger geographical context.
9. Stakeholder engagement is part of a larger process to internalize public values into complex water systems. This learning facilitates adaptation and evolution of water systems in response to climate change and changing societal needs.
10. Enable participants to learn about the process of making decisions as well as substantive aspects of the decisions themselves. Such knowledge helps users understand the organizational and political constraints on decision-making.
11. Effective engagement often occurs around scenario development. Scenarios are alternative storylines of the future. These stories should be plausible to a wide range of people.

8.4 Conclusions

The modern history of water development separated people from the water systems that supported their lives and livelihoods. Water science specialized in analytical research that advances knowledge by taking apart the whole to study its constituent parts. Analysis often comes at the expense of synthesis that is better suited to the study of complex systems like water. Water engineering worked on practical problems from a technical perspective, but ignored values, attitudes, and beliefs. The call for a new field of socio-hydrology in 2012 has begun to reverse these patterns by including interacting human-biophysical dynamics in the study of water resources (Sivapalan et al. 2012). Also significant, however, are the social relationships that surround socio-hydrology, water science, and water engineering. Building resilience for uncertain water futures will require a new social infrastructure for learning that unites scientists, water decision-makers, and the public in collective action to address the formidable challenge of managing water in the face of uncertainty.

References

Adger, W. Neil., Terry P. Hughes, Carl Folke, Stephen R. Carpenter, and Johan Rockström, 2005. "Social-Ecological Resilience to Coastal Disasters," *Science* 309: 1036–1039.

Akhmouch, Aziza and Delphone Clavreul. 2016. "Stakeholder Engagement for Inclusive Water Governance. Practicing What We Preach." *Water* 8(5): 204.

Anderson, W. C. 2004. "A History of Environmental Engineering in the United States." In *Water Resources and Environmental History*. Edited by J. R. Rogers, 103. Reston, VA: American Society of Civil Engineers.

Arnstein, Sherry R. 1969. "A Ladder of Citizen Participation." *Journal of the American Planning Association*, 35(4): 216–224. http://www.tandfonline.com/doi/abs/10.1080/01944366908977225. Accessed September 23 2017.

Bandura, Albert, and Richard H. Walters. 1963. *Social Learning and Personality Development*. New York: Holt, Rinehart & Winston.

Cash, David W., William C. Clark, Frank Alcock, Nancy M. Dickson, Noelle Eckley, David H. Guston, Jill Jäger, and Ronald B. Mitchell. 2003. "Knowledge Systems for Sustainable Development." *Proceedings of the*

National Academy of Sciences of the United States of America 10(14): 8086–8091.

Chang, Xiaolin, Xinglong Liu, and Wei Zhou. 2010. "Hydropower in China at Present and Its Further Development." *Energy* 35(11): 4400–4406.

Cook, Benjamin I., Toby R. Ault, and Jason E. Smerdon. 2015. "Unprecedented 21st Century Drought Risk in the American Southwest and Central Plains." *Science Advances,* 1(1): http://advances.sciencemag.org/content/1/1/e1400082. Accessed September 23 2017.

Dilling, Lisa, and Maria Carmon Lemos. 2011. "Creating Usable Science: Opportunities and Constraints for Climate Knowledge Use and Their Implications for Science Policy." *Global Environmental Change* 21(2): 680–689.

Disco, N., and Toussaint, B. 2014. "From Projects to Systems: The Emergence of a National Hydraulic Technocracy, 1990–1970." In *Two Centuries of Experience in Water Resources Management: A Dutch-US Retrospective,* edited by J. Lonnquest, B. Toussaint, J. Manous, Jr., and M. Ertsen, 155–200. Alexandria, VA: Rijkswaterstaat, Ministry of Infrastructure and the Environment and US Department of Defense, Army Corps of Engineers, Institute for Water Resources.

Folke, Carl, Thomas Hahn, Per Olsson, and Jon Norberg. 2005. "Adaptive Governance of Social-Ecological Systems." *Annual Review of Environment and Resources* 30: 441–473.

Global Reservoir and Dam Database. 2011. "GRanD Database." http://www.gwsp.org/products/grand-database/global-reservoir-and-dam-grand-database-project.html. Accessed August 18 2017.

Glynn, Pierre D., Alexey A. Voinov, Carl D. Shapiro, and Paul A. White. 2017. "From Data to Decisions: Processing Information, Biases, and Beliefs for Improved Manage of Natural Resources and Environments." *Earth's Future* 5(4): 356. https://doi.org/10.1002/2016EF000487. Accessed June 26 2017.

Gober, Patricia A., Graham E. Strickert, Douglas A. Clark, Kwok P. Chun, Diana Payton, and Kristin Bruce. 2015. "Divergent Perspectives on Water Security: Bridging the Policy Debate." *The Professional Geographer* 67(1): 62–71.

Hargrove, William L., Dewayne Johnson, Don Snethen, and Jan Middendorf. 2010. "From Dust Bowl to Mud Bowl: Sedimentation, conservation measures, and the future of reservoirs." *Journal of Soil and Water Conservation* 65 (1): 14A–17A.

Hassan, Fekri. 2011. *Water History for Our Time, Volume 2.* Paris, France: United Nations Educational, Scientific and Cultural Organization. http://unesdoc.unesco.org/images/0021/002108/210879e.pdf. Accessed June 26 2017.

Keller, L. Robin, Craig W. Kirkwood, and Nancy S. Jones. 2010. "Assessing Stakeholder Evaluation Concerns: An Application to the Central Arizona Water Resources System." *Systems Engineering* 13(1): 58–71.

Liu, Junguo, Chuanfu Zang, Shiying Tian, Jianguo Lui, Hong Yang, Shaofeng Jia, Liangzhi You, Bo Lui, and Miao Zhang. 2013. "Water Conservancy Projects in China: Achievements: Challenges and Way Forward." *Global Environmental Change* 23: 633–643.

Mahmood, K. 1987. "Reservoir Sedimentation: Impact, Extent, Mitigation" Technical Report WTP71. Washington, DC: World Bank. https://www.osti.gov/scitech/biblio/5564758. Accessed September 23 2017.

Molle, François. 2009. "Water and Society: New Problems Faced, New Skills Needed." *Irrigation and Drainage* 58: S205–S211.

National Research Council. 2007. *Evaluating Progress of the U.S. Climate Change Science Program: Methods and Preliminary Results*. Washington, DC: The National Academies Press. https://www.nap.edu/download/11934. Accessed September 23 2017.

Opperman, Jeffrey J., Gerald E. Galloway, Joseph Farione, Jeffrey F. Mount, Brian D. Richter, and Silvia Secchi. 2009. "Sustainable Floodplains Through Large-scale Reconnection to Rivers." *Science* 326(5959): 1487–1488.

Pahl-Wostl, Claudia. 2002. "Towards Sustainability in the Water Sector: The Importance of Human Actors and Processes of Social Learning." *Aquatic Sciences* 64(4): 394–411.

Pahl-Wostl, Claudia. 2009. "A Conceptual Framework for Analysing Adaptive Capacity and Multi-level Learning Processes in Resource Governance Regimes." *Global Environmental Change* 19(3): 354–365.

Pahl-Wostl, Claudia, and I. Borowski, 2007. "Methods for Participatory Water Resources Management–Preface." *Water Resources Management* 21: 1047–1048.

Pahl-Wostl, Claudia, Marc Craps, Art Dewulf, Erik Mostert, David Tabara, and Tharsi Tallieu, 2007. "Social Learning and Water Resource Management." *Ecology and Society* 12(2). 5 [online] http://www.ecologyandsociety.org/vol12/iss2/art5/. Accessed June 26 2017.

Pahl-Wostl, Claudia, Erik Mostert, and David Tabara. 2008. "The Growing Importance of Social Learning in Water Resource Management and Sustainability Science." *Ecology and Society* 13(1): 24.

Pahl-Wostl, Claudia, Georg Holtz, Britta Kastens, and Christian Knieper. 2010. "Analyzing Complex Water Governance Regimes: The Management and Transition Framework." *Environmental Science & Policy* 13(7): 571–581.

Pahl-Wostl, Claudia, Gert Becker, Christian Kneiper, and Jan Sendzmir. 2013. "How Multilevel Societal Learning Processes Facilitate Transformative Change: A Comparative Case Study Analysis on Flood Management," *Ecology and Society* 18(4): 58.

Parkins, John R., and Ross E. Mitchell. 2005. "Public Participation as Public Debate: A Deliberative Turn in Natural Resource Management." *Society & Natural Resources* 18(6): 529–540, https://doi.org/10.1080/08941920590947977.

Pearson Leonie J., Anthea Coogan, Wendy Proctor, and Timothy F. Smith. 2010. "A Sustainable Decision Support Framework for Urban Water Management." *Water Resources Management* 24: 363–376.

Quay, Ray, Kelli L. Larson, and Dave D. White. 2013. "Enhancing Water Sustainability Through University Policy Collaborations: Experiences and Lessons from Researchers and Decision Makers." *Water Resources Impact* 5(2): 17–19.

Rodriguez-Iturbe, I. 2000. "Ecohydrology: A Hydrologic Perspective of Climate-soil-vegetation Dynamics." *Water Resources Research* 36(1): 3–9.

Sivapalan, Murugesu, Hubert H. G. Savenije, and Gunter Blösch. 2012. "Socio-hydrology: A New Science of People and Water," *Hydrological Processes* 26(2): 1270–1276.

Strickert, Graham E. H. and Lori Bradford. 2015. "Of Research Pings and Ping-Pong Balls: The Use of Forum Theater for Engaged Water Security Research." *International Journal of Qualitative Methods* 15: 1–14. https://doi.org/10.1177/1609406915621409.

Tippett, J., B. Searle, C. Pahl-Wostl, and Y. Rees. 2005. "Social Learning in Public Participation in River Basin Management—Early Findings from HarmoniCOP European Case Studies." *Environmental Science & Policy* 8(3): 287–299.

US Army Corps of Engineers. 2017. "The Army Corps of Engineers: A Brief History." http://www.usace.army.mil/About/History/Brief-History-of-the-Corps/Beginnings/. Accessed August 18 2017.

US Environmental Protection Agency. 2017. "What is Green Infrastructure?" https://www.epa.gov/green-infrastructure/what-green-infrastructure. Accessed August 18 2017.

von der Porten, Suzanne. 2014. "How Collaborative Approaches to Environmental Problem Solving View Indigenous Peoples: A Systematic Review. " *Society and Natural Resources* 27: 1040–1056.

Vörösmarty Charles J., Peter B. McIntyre, Mark O. Gessner, David Dudgeon, Alexander Prusevich, Pamela Green, Stanley Glidden et al. 2010. "Global Threats to Water Security." *Nature* 467(7315): 555–561.

Wade, Rebecca, and Neil McLean. 2014. "Multiple Benefits of Green Infrastructure." In *Water Resources in the Built Environment, Management Issues and Solutions*. Edited by Colin A. Booth and Susanne M. Charlesworth, 319–335. Chichester: John Wiley & Sons Ltd. https://doi.org/10.1002/9781118809167.ch24.

White, R. 2001. *Evacuation of Sediments from Reservoirs*. London: Thomas Telford. https://doi.org/10.1680/eosfr.29538.

White, Dave D., Amber Y. Wutich, Kelli L. Larson, and T. Lant. 2015. "Water Management Decision Makers' Evaluations of Uncertainty in a Decision Support System: The Case of WaterSim in the Decision Theater." *Journal of Environmental Planning and Management* 58(4): 616–630.

Winiwarter, Verena, Gertrud Haidvogl, Severin Hohensinner, Friedrich Hauer, and Michael Bürkner. 2016. "The Long-Term Evolution of Urban Waters and Their Nineteenth Century Transformation in European Cities. A Comparative Environmental History." *Water History* 8: 209–233. https://doi.org/10.1007/s12685-016-0172-z. http://onlinelibrary.wiley.com/doi/10.1002/wrcr.20452/pdf. Accessed June 26 2017.

Wisser, Dominik. Steve Frolking, Stephan Hagen and Marc F. P. Bierkens. 2013. "Beyond Peak Reservoir Storage? A global Estimate of Declining Water Storage Capacity in Large Reservoirs." *Water Resources Research* 49: 5732–5739.

Yin, Hongfu., and Changan Li. 2001. "Human Impact on Floods and Flood Disasters on the Yangtze River." *Geomorphology* 41: 105–109.

9

Strategies for Resilience

9.1 Introduction

For thousands of years, humans have sought to control water for transportation, food and energy production, human consumption, flood protection, and economic development. In recent times, the assumption of stationarity guided this work, and emphasis was on designing infrastructure, operational rules, allocation systems, and governance mechanisms to put water to optimum use to meet human needs. The uncertainties of climate change raise the specter that water systems cannot be easily controlled in a non-stationary future, and new strategies are needed for long-term water planning.

Today's water systems are already stressed from overuse, governance failures, loss of storage capacity, rapid population growth, planning failures, changing lifestyles, and meat-eating diets. They are vulnerable to climate change impacts because of human mismanagement and failure to anticipate a future that may be different from the past. Response to recent Hurricanes Harvey and Irma is a case in point. Although the 2017 storms were powerful, the root cause of severe damage was development patterns in Florida and Houston. Developers in Florida constructed new

P. Gober, *Building Resilience for Uncertain Water Futures*,
https://doi.org/10.1007/978-3-319-71234-5_9

condominiums adjacent to the coastline and in wetlands that had provided natural protection in flood events (Alverez 2017). Development of low-lying neighborhoods in Houston reflects the same go-go attitude that exposed thousands of residents to the vagaries of Hurricane Harvey after severe flooding in 2015 and 2016 (Whoriskey and Sullivan 2017). Homeowners today lament the loss in housing values and neighborhood life stemming from Harvey's damage. City leaders worry about the potential for vacant lots in their jurisdictions if owners do not rebuild. Building contractors line up to construct replacement homes up to seven feet above street level to protect them from future flooding.

It is increasingly clear that these episodes are not the result of rare and unexpected natural events, but symptoms of human failures to interpret the underlying pattern of overuse, overbuilding, contamination, inequality, careless regulation, and environmental neglect. This pattern has increased vulnerability to climatic events, despite impressive new technology to predict storms and logistic expertise to evacuate and respond to immediate emergencies.

The predict-and-control paradigm of modern water management is not working well when flood events occur more often than they should, water tables fall unexpectedly, and dams function for 75 rather than 150 years. Predict-and-control management systems are unable to accommodate climatic uncertainties associated with a non-stationary future. This book offers a new way of addressing water problems that embraces uncertainty, acknowledges the central role of human agency in water systems, seeks to design a more sustainable future, recognizes the trade-offs embedded in water decisions, and builds institutions to promote social learning. Wicked water problems are managed, not resolved.

This model of water management implies a changing role for water scientists. There are no easy technological fixes or scientific answers to wicked water problems. They require instead the capacity to focus on the long term and move away from asking scientists to predict the most likely future. They rely on community partners to say what kind of future they want and what policy decisions they are willing to accept to get there. This model of water management favors new types of scientific practice, including building water resource models to explore the future, deeper understanding of the consequences of policy decisions, and collaboration with stakeholders to learn what important questions to ask. Living with uncertainty also requires accepting mistakes and learning from them, seeing policies as experiments, and relentless monitoring of water systems. Resilient water systems are able to adjust to inevitable shocks from climate change and other sources and maintain essential functions.

9.2 How to Adapt Water Systems for Uncertain Water Futures

9.2.1 Emphasize the Human Dimensions

Current water research favors climate science, hydrology, and ecology at the expense of markets, policy, land use, behaviors, lifestyles, and values. It is not possible to understand the history, status, or future of water systems without accounting for human agency (Sivapalan et al. 2012). Damage from water-related hazards, resource shortages, and inadequate sanitation are less science problems than failures of governance and practice. Phoenix area water managers are far less interested in the technical details of climate science and more interested in the impacts of a range of policy levers on future supplies and demand for water. They weigh scientific uncertainty against political, social, and economic uncertainty to make decisions about the future. There is a need for better knowledge about the consequences of policy decisions and technology on behavior, lifestyles, water use, and future outcomes. These are primarily social science and policy research questions for a field that is heavily weighed toward science and engineering practice.

9.2.2 Reduce Vulnerabilities

There is a great deal of uncertainty in current global climate models, and the best minds in the field anticipate that uncertainty will increase in the short to mid-term (Trenberth 2010). Top-down methods of climate adaptation have limited ability to anticipate future climate change impacts, not to mention society's reaction to them. Bottom-up strategies seek to reduce the vulnerabilities that already exist, starting with improving governance and allocation, protecting water sources from pollution and salinization, restricting settlement in flood-prone areas and incentivizing people to move out of these areas, improving water efficiency (in industry, agriculture, and domestic use), and increasing awareness-raising and risk-spreading instruments (insurance).

The social science community has developed deep knowledge about the root causes of the loss of life and property from extreme events. It has, however, focused on place-based assessments that are difficult to translate into national policy initiatives. Water resource modeling has also focused on case studies without adequate efforts at synthesizing results for a broader understanding of the impacts of policies on vulnerability reduction. Both fields are in need of integrated research that would focus on exploratory modeling for vulnerability reduction, DMUU for decision support, and social learning for collective action.

9.2.3 Collaborate with Related Sectors

It is increasingly risky to ignore vulnerabilities in complex interdependent systems such as water, energy, land, and food. National governments, private sector actors, local planning authorities, and regional governments have a shared interest in understanding how interdependencies and feedbacks lead to cascading failures and shocks. The stovepipes of government sometimes limit cross-sector vulnerability assessments of existing infrastructure and the implications of drought and increasing hydrological variability (Moss et al. 2016). There should be stronger collaboration among agencies involved in the WELF nexus to model and manage the dynamics of evolving connected infrastructure and to plan for potential shocks. Institutionalized collaboration ensures that land use decisions incorporate water requirements, energy development considers the constraints of drought and water shortages, and the indirect consequences of governance policy (e.g., support for biofuel development) include not only energy but also water and environmental factors.

9.2.4 Anticipate Surprise

The military knows how to anticipate surprise as a critical strategy (Grabo 2002). It spends a fair bit of time collecting information to anticipate future enemy actions. Whether or not there is a crisis or imminent threat on the horizon, the mission of the intelligence community is to examine continually and report periodically. While water problems are not generally military actions, they do require careful monitoring, watchfulness, warnings, hypothesizing about "what if" events, and a mindset quite different from predict and plan. The water sector may benefit from what the military has learned about the kinds of problems that trigger surprise and the thinking needed to avoid them in the future. Repeated surprises occur when there is inadequate perception of emerging threat, particularly those of low probability and great danger, breakdown in communication among agencies involved in the threat, and failure to heed minority viewpoints. There needs to be an attitude of continuous watchfulness, attention to worst-case scenarios, ongoing analysis of indicators, recognition of the limits of current knowledge, presumption of surprise, ability to see connections, and need to make decisions before all the facts are in. These tend not to be traits associated with the water sector historically, but better aligned to the new challenge of climatic uncertainty.

9.2.5 Learn to Live with Uncertainty

The scientific tradition is to reduce uncertainty as a mechanism to provide better predictions of the future. The essence of non-stationarity is that the future is inherently uncertain, and we need to learn to live with it and make better decisions in the face of it. Efforts to maximize efficiency (e.g., Thai supply chains) can lead to unexpected and undesirable outcomes in a complex resource system and global economy. In making systems more efficient, we sacrifice resilience—the capacity to bounce back after an emergency. Water systems require contingency plans, excess capacity, back-up suppliers, and precautionary actions to move people from dangerous places to manage climatic and other uncertainties. The private sector, as we learned in Chap. 4, has revamped its supply chains to avoid the catastrophic consequences of single suppliers and failure to make contingency plans. These plans are the stuff of DMUU.

9.2.6 Embrace Values and Feelings and Trade-offs

Water managers have always incorporated political, economic, and legislative uncertainty into their decision processes. Stakeholders add to that mix and increase the range of concerns that is considered. Public utilities, largely shielded from the public, now balance the needs of water's many constituents with competing interests. Recent drought conditions in California forced regulators to choose among the needs of farmers, recreationalists, industry, urban residents, energy production, and the environment. Dam management in Canada weighs demand for hydropower, environmental flows, and the needs of Indigenous peoples to practice their traditional livelihoods. Phoenix area water managers weigh the economic benefits of population growth against the need for conservation and expensive infrastructure to avoid long-term water shortages. Modeling efforts that highlight trade-offs enhance deeper public understanding of what it will take to achieve the future that they want. Values are grist for assigning priority to one set of concerns over another. Values incorporate how people feel about a problem. Feelings, along with analysis, play a role in water decisions.

9.2.7 Use DMUU Strategies

There are no ideal solutions for wicked problems, but rather strategies for handling uncertainty, complexity, and political conflict. DMUU changes the conversation from what is the best solution to what kind of future we want and what are the best strategies for achieving it. DMUU looks for robust solutions that work well regardless of future climate and hydrological outcomes. Science is critical to DMUU, but not in the sense that it identifies the single best path forward. It illuminates the range of alternative futures, monitors system changes to ensure they are moving in desirable directions, assesses whether policies are working, and identifies the need to change course if the need arises. Public engagement ensures that a variety of viewpoints inform water decisions. In a recent hearing about whether to increase water rates in Tempe, Arizona, few citizens showed up to express an opinion about increasing water rates (to raise revenue and foster conservation). The only public input was from large users who opposed the rate increases and favored the current flat rate structure. The city concluded that there is no public support for rate increases, and the rates remained constant.

9.2.8 Emphasize Co-benefit Solutions

Urban climate adaptation plans emphasize strategies that create value in the process of lessening climate change impacts. Rotterdam's investments in green infrastructure improve the city's livability, walkability, and potential to attract tourism. London aspires to export its climate adaptation expertise, and Singapore's new barrage is part tourist attraction and public space in addition to a means of managing higher precipitation variability. Efforts in urban water conservation save energy and treatment costs and enable future growth. Increasingly, cities reframe conservation programs from efforts to save water to programs that provide a new water source.

9.2.9 Infuse Social Learning into Water Planning

Embedding water science and decision-making with social learning requires a radical departure in the way scientists, water managers, and citizens have traditionally seen water problems and their solutions. Many scientists still work on specialized research projects. Water managers often seek input from consultants to solve narrow technical problems (e.g., infrastructure design, flood mapping, integrated water resource modeling, water testing, water planning, water rights acquisition, irrigation practice, and reservoir operations) but less so from independent scientists with research programs that potentially dovetail with agency responsibilities. Close relationships with independent scientists as partners in advanced modeling, systematic monitoring, and assessment are not so common but are increasing in the field. Neither scientists nor water managers have vast experience in public engagement, with many activities focused on educating the public or calming public worries at the bottom end of Arnstein's Ladder of Civic Participation. Social learning is a mechanism for knitting science, practice, and the public on issues that address problems raised in the early chapters and resolve (or at least agree to move forward in the face of) inherent controversies associated with competing interests and different knowledge systems articulated in the latter chapters. The marriage of socio-hydrology and DMUU offers a path forward for conceptualizing water problems (as failures of governance) and building resilience in the water sector.

9.3 Conclusions

A sea change has occurred in water science and management during the past 15 years. There is now acceptance of the role of social science in understanding water problems and in designing new schemes for governance, social learning, policy, and practice. To date, however, social scientists have focused on idiographic case studies with relatively little effort to produce knowledge at scale for global impact. Elinor Ostrom's (1990) work on governing the commons is an important exception and much-needed model for social science contributions to resource management in general and water science and practice in particular. Social science introduces new perspectives on water science, including the idea that uncertainty is an inescapable fact of life. DMUU provides a framework for making decisions despite uncertainty.

This book does not use the term "paradigm change" loosely. The emphasis on infrastructure, control, optimization, and prediction brought us to a point where it is hard to see beyond the past. Today's water system problems were inevitable outcomes of overuse, economic growth, changing lifestyles, increasing inequality, and lack of foresight. Climate change promises to complicate all this and to add to uncertainties about the future. A new paradigm of water management, one better equipped to deal with an uncertain future, incorporates competing values, and treats water as a limited resource. It focuses on multiple scenarios of the future, policies to reduce vulnerabilities, and building resilience to an uncertain future in the water sector.

References

Alverez, Lisette. 2017. "Despite Rising Seas and Bigger Storms, Florida's Land Rush Endures." *New York Times*, September 18 2017.

Grabo, Cynthia M. 2002. "Anticipating Surprise: Analysis for Strategic Warning." Washington, DC: US Defense Intelligence Agency. http://www.ni-u.edu/ni_press/pdf/Anticipating_Surprise_Analysis.pdf. Accessed September 24 2017.

Moss, Richard H., Karen Fisher-Vanden, Alison Delgado, Scott Backhaus, Christopher L. Barrett, Bhaduri, Budhendra, Ian P. Kracunas, Patrick M. Reed, Jennie S. Rice, Ian Sue Wing, and Claudia Tebaldi. 2016. "Understanding Dynamics and Resilience in Complex Interdependent Systems: Prospects for a Multi-Model Framework and Community of Practice." Report of a workshop held under the auspices of the US Global Change Research Program. https://climatemodeling.science.energy.gov/publications/understanding-dynamics-and-resilience-complex-interdependent-systems. Accessed September 12 2017.

Ostrom, Elinor. 1990. *Governing the Commons: The Evolution of Institutions for Collective Action*. Cambridge: Cambridge University Press.

Sivapalan, Murugesu, Hubert H. G. Savenije, and Gunter Blösch. 2012. "Socio-hydrology: A New Science of People and Water." *Hydrological Processes* 26(2): 1270–1276.

Trenberth, Kevin. 2010. "More Knowledge less Certainty," *Nature Reports Climate Change* 4(February): 20–21.

Whoriskey, Peter and Patricia Sullivan. 2017. "Residents of Flood Weary Houston Neighborhoods Wonder if Rebuilding Is Worth It." *Washington Post*, September 14 2017.

Index[1]

[1] Note: Page numbers followed by 'n' refer to notes.

P. Gober, *Building Resilience for Uncertain Water Futures*,
https://doi.org/10.1007/978-3-319-71234-5

The manufacturer's authorised representative in the EU is Springer Nature Customer Service Centre GmbH, Europaplatz 3, 69115 Heidelberg, Germany. If you have any concerns regarding our products, please contact ProductSafety@springernature.com

Printed and bound by CPI Group (UK) Ltd, Croydon, CR0 4YY
30/09/2026
02225235-0002